FOREST, FIRE CONTROL AND PROTECTION

FOREST, FIRE CONTROL AND PROTECTION

Reepunjaya Singh

Oxford Book Company
Jaipur ▪ New Delhi

Published By

Oxford Book Company

Regd. Off. "Bony Residency", Gate No. 2
Opp. Tilak Public School, Vishweswariya Nagar,
Gopalpura Road, Jaipur - 302018, India
Ph: 0141 - 2761280, Telefax: 2761381
email - oxfordbookjpr@gmail.com
website - oxfordbookcompany.com

Delhi Office: 102, 1st Floor, Satyam House,
4327/3, Ansari Road, Darya Ganj,
New Delhi - 110002, India
Ph: 011-45652440

ributed By
·A INTERNATIONAL PRESS LLP
E-mail: keyainternationalpressllp@gmail.com

Published in India, 2025

ISBN : 9789391096915

Typeset by Shivangi Computers, Jaipur

Printed and bound in India by Thomson Press Ltd.,
New Delhi

Preface

Forest Fire management is a branch of forestry concerned with the overall administrative, economic, legal and social aspects and with the essentially scientific and technical aspects, especially silviculture, protection, and forest regulation. This includes management for aesthetics, fish, recreation, urban values, water, wilderness, wildl[illegible] wood products, forest genetic resources and other forest resource v[illegible] Management can be based on conservation, economics, or a m[illegible] of the two. Techniques include timber extraction, planting and re[illegible] of various species, cutting roads and pathways through for[illegible] and preventing fire. In the economy of man and of nature forests are of direct and indirect value, the former chiefly through the produce which they yield and the latter through the influence which they exercise upon climate, the regulation of moisture, the stability of the soil, the healthiness and beauty of a country and allied subjects.

Thousands of hectares around the globe destroyed by forest fires every year causing tragic loss of houses, properties, lives, fauna and flora. Forest fires are a great menace to ecologically healthy grown forests and protection of the environment. This problem has been the research interest for years, and there are a number of solutions available to resolve this problem.

A forest fire is unenclosed and freely spreading combustion that consumes the natural fuels of a forest—i.e., duff, grass, weeds, brush, and trees. Forest fires occur in three principal forms, the distinctions depending essentially on their mode of spread and their position in relation to the ground surface. Surface fires burn surface litter, other loose debris of the forest floor, and small vegetation; a surface fire may, and often does, burn taller vegetation and tree crowns as it progresses. Crown fires advance through the tops of trees or shrubs more or less independently of the surface fire and are the fastest spreading of all forest fires. Ground fires consume the organic material beneath the surface litter of the forest floor; ground fires are the least spectacular and the slowest-moving, but they are often the most destructive of all forest fires and also the most difficult to control.

This book come from well known scientists/researchers, teachers and academicians and represents a collection of well-written articles on the topics covered.

— Reepunjaya Singh

Contents

CHAPTER 1

INTRODUCTION TO FOREST

Definition and Characteristics of Forest

A forest is an area of land dominated by trees. Hundreds of definitions of forest are used throughout the world, incorporating factors such as tree density, tree height, land use, legal standing and ecological function. The Food and Agriculture Organization defines a forest as land spanning more than 0.5 hectares with trees higher than 5 meters and a canopy cover of more than 10 percent, or trees able to reach these thresholds in situ. It does not include land that is predominantly under agricultural or urban land use. Using this definition FRA 2020 found that forests covered 4.06 billion hectares or approximately 31 percent of the global land area in 2020 but are not equally distributed around the globe.

Forests are the dominant terrestrial ecosystem of Earth, and are distributed around the globe. More than half of the world's forests are found in only five countries (Brazil, Canada, China, Russian Federation and United States of America). The largest part of the forest (45 percent) is found in the tropical domain, followed by the boreal, temperate and subtropical domains.

Forests account for 75% of the gross primary production of the Earth's biosphere, and contain 80% of the Earth's plant biomass. Net primary production is estimated at 21.9 gigatonnes carbon per year for tropical forests, 8.1 for temperate forests, and 2.6 for boreal forests.

Forests at different latitudes and elevations, and with different precipitation and evapotranspiration form distinctly different biomes: boreal forests around the North Pole, tropical moist forests and tropical dry forests around the Equator, and temperate forests at the middle latitudes. Higher elevation areas tend to support forests similar to those at higher latitudes, and amount of precipitation also affects forest composition.

Almost half the forest area (49 percent) is relatively intact, while 9 percent is found in fragments with little or no connectivity. Tropical rainforests and boreal coniferous forests are the least fragmented, whereas subtropical dry forest and temperate oceanic forests are among the most fragmented. Roughly 80 percent of the world's forest area is found in patches larger than 1 million hectares. The remaining 20 percent is located in more than 34 million patches across the world – the vast majority less than 1 000 hectares in size.

Human society and forests influence each other in both positive and negative ways. Forests provide ecosystem services to humans and serve as tourist attractions. Forests can also affect people's health. Human activities, including unsustainable use of forest resources, can negatively affect forest ecosystems.

Definition

Although the word *forest* is commonly used, there is no universally recognised precise definition, with more than 800 definitions of forest used around the world. Although a forest is usually defined by the presence of trees, under many definitions an area completely lacking trees may still be considered a forest if it grew trees in the past, will grow trees in the future, or was legally designated as a forest regardless of vegetation type.

There are three broad categories of forest definitions in use: administrative, land use, and land cover. Administrative definitions are based primarily upon the legal designations of land, and commonly bear little relationship to the vegetation growing on the land: land that is legally designated as a forest is defined as a forest even if no trees are growing on it. Land use definitions are based upon the primary purpose that the land serves. For example, a forest may be defined as any land that is used primarily for production of timber. Under such a land use definition, cleared roads or infrastructure within an area used for forestry, or areas within the region that have been cleared by harvesting, disease or fire are still considered forests even if they contain no trees. Land cover definitions define forests based upon the type and density of vegetation growing on the land. Such definitions typically define a forest as an area growing trees above some threshold. These thresholds are typically the number of trees per area (density), the area of ground under the tree canopy (canopy cover) or the section of land that is occupied by the cross-section of tree trunks (basal area). Under such land cover definitions, an area of land can only be known as forest if it is growing trees. Areas that fail to meet the land cover definition may be still included under while immature trees are establishing if they are expected to meet the definition at maturity.

Under land use definitions, there is considerable variation on where the cutoff points are between a forest, woodland, and savanna. Under some definitions, forests require very high levels of tree canopy cover, from 60% to 100%, excluding savannas and woodlands in which trees have a lower canopy cover. Other definitions consider savannas to be a type of forest, and include all areas with tree canopies over 10%.

Some areas covered with trees are legally defined as agricultural areas, e.g. Norway spruce plantations in Austrian forest law when the trees are being grown as Christmas trees and below a certain height.

Evolutionary history

The first known forests on Earth arose in the Late Devonian (approximately 380 million years ago), with the evolution of *Archaeopteris*. *Archaeopteris* was a plant that was both tree-like and fern-like, growing to 10 metres (33 ft) in height. *Archaeopteris* quickly spread throughout the world, from the equator to subpolar latitudes. *Archaeopteris* formed the first forest by being the first known species to cast shade due to its fronds and forming soil from its roots. *Archaeopteris* was deciduous, dropping its fronds onto the forest floor. The shade, soil, and forest duff from the dropped fronds created the first forest. The shed organic matter altered the freshwater environment, slowing it down and providing food. This promoted freshwater fish.

Ecology

Forest ecology is the scientific study of the interrelated patterns, processes, flora, fauna and ecosystems in forests. The management of forests is known as forestry, silviculture, and forest management. A forest ecosystem is a natural woodland unit consisting of all plants, animals and micro-organisms (Biotic components) in that area functioning together with all of the non-living physical (abiotic) factors of the environment. The forest ecosystem is very important.

Figure: Temperate rainforest in Tasmania's Hellyer Gorge

Forests account for 75% of the gross primary productivity of the Earth's biosphere, and contain 80% of the Earth's plant biomass.

The world's forests contain about 606 gigatonnes of living biomass (above- and below-ground) and 59 gigatonnes of dead wood. The total biomass has decreased slightly since 1990 but biomass per unit area has increased.

Forest ecosystems can be found in all regions capable of sustaining tree growth, at altitudes up to the tree line, except where natural fire frequency or other disturbance is too high, or where the environment has been altered by human activity.

The latitudes 10° north and south of the equator are mostly covered in tropical rainforest, and the latitudes between 53°N and 67°N have boreal forest. As a general rule, forests dominated by angiosperms (*broadleaf forests*) are more species-rich than those dominated by gymnosperms (*conifer*, *montane*, or *needleleaf forests*), although exceptions exist.

Forests sometimes contain many tree species within a small area (as in tropical rain and temperate deciduous forests), or relatively few species over large areas (e.g., taiga and arid montane coniferous forests). Forests are often home to many animal and plant species, and biomass per unit area is high compared to other vegetation communities. Much of this biomass occurs below ground in the root systems and as partially decomposed plant detritus. The woody component of a forest contains lignin, which is relatively slow to decompose compared with other organic materials such as cellulose or carbohydrate.

The biodiversity of forests varies considerably according to factors such as forest type, geography, climate and soils – in addition to human use. Most forest habitats in temperate regions support relatively few animal and plant species and species that tend to have large geographical distributions, while the montane forests of Africa, South America and Southeast Asia and lowland forests of Australia, coastal Brazil, the Caribbean islands Central America and insular Southeast Asia have many species with small geographical distributions. Areas with dense human populations and intense agricultural land use, such as Europe, parts of Bangladesh, China, India and North America, are less intact in terms of their biodiversity. Northern Africa, southern Australia, coastal Brazil, Madagascar and South Africa, are also identified as areas with striking losses in biodiversity intactness.

Components

Figure: Even, dense old-growth stand of beech trees (Fagus sylvatica) prepared to be regenerated by their saplings in the understory, in the Brussels part of the Sonian Forest.

A forest consists of many components that can be broadly divided into two categories that are biotic (living) and abiotic (non-living) components. The living parts include trees, shrubs, vines, grasses and other herbaceous (non-woody) plants, mosses, algae, fungi, insects, mammals, birds, reptiles, amphibians, and microorganisms living on the plants and animals and in the soil.

Layers

Figure: Spiny forest at Ifaty, Madagascar, featuring various Adansonia (baobab) species, Alluaudia procera (Madagascar ocotillo) and other vegetation

A forest is made up of many layers. The main layers of all forest types are the forest floor, the understory and the canopy. The emergent layer exists in tropical rainforests. Each layer has a different set of plants and animals depending upon the availability of sunlight, moisture and food.

- **Forest floor** contains decomposing leaves, animal droppings, and dead trees. Decay on the forest floor forms new soil and provides nutrients to the plants. The forest floor supports ferns, grasses, mushroom and tree seedlings.
- **Understory** is made up of bushes, shrubs, and young trees that are adapted to living in the shades of the canopy.
- **Canopy** is formed by the mass of intertwined branches, twigs and leaves of the mature trees. The crowns of the dominant trees receive most of the sunlight. This is the most productive part of the trees where maximum food is produced. The canopy forms a shady, protective "umbrella" over the rest of the forest.
- **Emergent layer** exists in the tropical rain forest and is composed of a few scattered trees that tower over the canopy.

However, in botany and many countries (Germany, Poland, etc.), a different classification of forest vegetation structure is often used: tree, shrub, herb and moss layers.

Types

Figure: A dry sclerophyll forest in Sydney, which is dominated by eucalyptus trees.

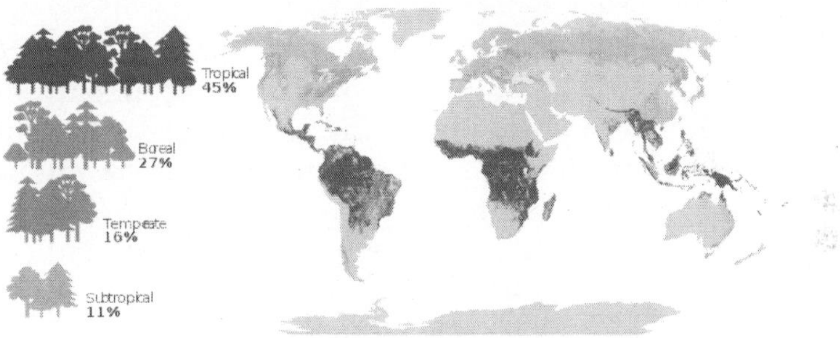

Figure: Proportion and distribution of global forest area by climatic domain, 2020

Forests can be classified in different ways and to different degrees of specificity. One such way is in terms of the biome in which they exist, combined with leaf longevity of the dominant species (whether they are evergreen or deciduous). Another distinction is whether the forests are composed predominantly of broadleaf trees, coniferous (needle-leaved) trees, or mixed.

- Boreal forests occupy the subarctic zone and are generally evergreen and coniferous.
- Temperate zones support both broadleaf deciduous forests (*e.g.*, temperate deciduous forest) and evergreen coniferous forests

(*e.g.*, temperate coniferous forests and temperate rainforests). Warm temperate zones support broadleaf evergreen forests, including laurel forests.

- Tropical and subtropical forests include tropical and subtropical moist forests, tropical and subtropical dry forests, and tropical and subtropical coniferous forests.
- Physiognomy classifies forests based on their overall physical structure or developmental stage (e.g. old growth vs. second growth).
- Forests can also be classified more specifically based on the climate and the dominant tree species present, resulting in numerous different forest types (e.g., Ponderosa pine/Douglas-fir forest).

The number of trees in the world, according to a 2015 estimate, is 3 trillion, of which 1.4 trillion are in the tropics or sub-tropics, 0.6 trillion in the temperate zones, and 0.7 trillion in the coniferous boreal forests. The estimate is about eight times higher than previous estimates, and is based on tree densities measured on over 400,000 plots. It remains subject to a wide margin of error, not least because the samples are mainly from Europe and North America.

Forests can also be classified according to the amount of human alteration. Old-growth forest contains mainly natural patterns of biodiversity in established seral patterns, and they contain mainly species native to the region and habitat. In contrast, secondary forest is regrowing forest following timber harvest and may contain species originally from other regions or habitats.

Different global forest classification systems have been proposed, but none has gained universal acceptance. UNEP-WCMC's forest category classification system is a simplification of other more complex systems (e.g. UNESCO's forest and woodland 'subformations'). This system divides the world's forests into 26 major types, which reflect climatic zones as well as the principal types of trees. These 26 major types can be reclassified into 6 broader categories: temperate needleleaf; temperate broadleaf and mixed; tropical moist; tropical dry; sparse trees and parkland; and forest plantations. Each category is described as a separate section below.

Temperate needleleaf

Temperate needleleaf forests mostly occupy the higher latitude regions of the Northern Hemisphere, as well as high altitude zones and some warm temperate areas, especially on nutrient-poor or otherwise unfavourable soils. These forests are composed entirely, or nearly so, of coniferous species (Coniferophyta). In the Northern Hemisphere pines *Pinus*, spruces *Picea*, larches *Larix*, firs *Abies*, Douglas firs *Pseudotsuga* and hemlocks *Tsuga*, make up the canopy, but other taxa are also important. In the Southern Hemisphere, most coniferous trees (members of the Araucariaceae and Podocarpaceae) occur in mixtures with broadleaf species, and are classed as broadleaf and mixed forests.

Temperate broadleaf and mixed

Figure: Broadleaf forest in Bhutan

Temperate broadleaf and mixed forests include a substantial component of trees in the Anthophyta. They are generally characteristic of the warmer temperate latitudes, but extend to cool temperate ones, particularly in the southern hemisphere. They include such forest types as the mixed deciduous forests of the United States and their counterparts in China and Japan, the broadleaf evergreen rainforests of Japan, Chile and Tasmania, the sclerophyllous forests of Australia, central Chile, the Mediterranean and California, and the southern beech Nothofagus forests of Chile and New Zealand.

Tropical moist

There are many different types of tropical moist forests, with lowland evergreen broad leaf tropical rainforests, for example várzea and igapó forests and the terra firma forests of the Amazon Basin; the peat swamp forests, dipterocarp forests of Southeast Asia; and the high forests of the Congo Basin. Seasonal tropical forests, perhaps the best description for the colloquial term "jungle", typically range from the rainforest zone 10 degrees north or south of the equator, to the Tropic of Cancer and Tropic of Capricorn. Forests located on mountains are also included in this category, divided largely into upper and lower montane formations on the basis of the variation of physiognomy corresponding to changes in altitude.

Tropical dry

Tropical dry forests are characteristic of areas in the tropics affected by seasonal drought. The seasonality of rainfall is usually reflected in the deciduousness of the forest canopy, with most trees being leafless for several months of the year. However, under some conditions, e.g. less fertile soils or less predictable drought regimes, the proportion of evergreen species increases and the forests are characterised as "sclerophyllous". Thorn forest, a dense forest of low stature with a high frequency of thorny or spiny species, is found where drought is prolonged, and especially where grazing animals are plentiful. On very poor soils, and especially where fire or herbivory are recurrent phenomena, savannas develop.

Sparse trees and parkland

Figure: Taiga forest near Saranpaul in the northeast Ural Mountains, Khanty–Mansia, Russia. Trees include Picea obovata *(dominant on right bank),* Larix sibirica, Pinus sibirica, *and* Betula pendula.

Sparse trees and savanna are forests with lower canopy cover of trees. They occur principally in areas of transition from forested to non-forested landscapes. The two major zones in which these ecosystems occur are in the boreal region and in the seasonally dry tropics. At high latitudes, north of the main zone of boreal forest, growing conditions are not adequate to maintain a continuous closed forest cover, so tree cover is both sparse and discontinuous. This vegetation is variously called open taiga, open lichen woodland, and forest tundra. A savanna is a mixed woodland grassland ecosystem characterized by the trees being sufficiently widely spaced so that the canopy does not close. The open canopy allows sufficient light to reach the ground to support an unbroken herbaceous layer consisting primarily of grasses. Savannas maintain an open canopy despite a high tree density.

Forest plantations

Forest plantations are generally intended for the production of timber and pulpwood. Commonly mono-specific, planted with even spacing between the trees, and intensively managed, these forests are not generally important as habitat for native biodiversity. However, they can be managed in ways that enhance their biodiversity protection functions and they can provide ecosystem services such as maintaining nutrient capital, protecting watersheds and soil structure, and storing carbon.

Forest area

The net loss of forest area has decreased substantially since 1990, but the world is not on track to meet the target of the United Nations Strategic Plan for Forests to increase forest area by 3 percent by 2030.

While deforestation is taking place in some areas, new forests are being established through natural expansion or deliberate efforts in others. As a result, the net loss of forest area is less than the rate of deforestation and it too is decreasing: from 7.8 million hectares per year in the 1990s to 4.7 million hectares per year during 2010– 2020. In absolute terms, the global forest area decreased by 178 million hectares between 1990 and 2020, which is an area about the size of Libya.

Surrounding issues

Forests have an enormously important role to play in our ecosystem. Not only are they essential for humans to breathe at all, they also serve as homes for millions of people, and billions depend on forests in some way. Likewise, a large proportion of the world's animal species live in forests. That's why we absolutely must protect them. Forest ecology helps to understand life in the forest. It shows how living organisms behave, live and survive. Furthermore, forest ecology attempts to provide an understanding of the thriving process of forests. Similarly, many forests are also used for economic purposes such as fuel and wood products. Therefore, understanding how forests are managed is also a goal of forest ecology.[

Societal significance

Figure: Burned forest on Thasos

Forests provide a diversity of ecosystem services including:

- converting carbon dioxide into oxygen and biomass. A full-grown tree produces about 100 kg of net oxygen per year.
- acting as a carbon sink. Therefore, they are necessary to mitigate climate change. According to the Special Report on Global Warming of 1.5 °C of the Intergovernmental Panel on Climate Change, to avoid temperature rise by more than 1.5 degrees above pre-industrial levels, there will need to be an increase in global forest cover equal to the land area of Canada (10 million km^2), by the year 2050.
- aiding in regulating climate. For example, a research from 2017, show that forests induce rainfall. If the forest is cut, it can lead to drought.
- purifying water.
- mitigating natural hazards such as floods.
- serving as a genetic reserve.
- serving as a source of lumber and as recreational areas.

- Serving as a source of woodlands and trees for millions of people depending almost entirely on forests for subsistence for their essential fuelwood, food and fodder needs.

Some researchers state that forests do not only provide benefits, but can in certain cases also incur costs to humans. Forests may impose an economic burden, diminish the enjoyment of natural areas, reduce the food producing capacity of grazing land and cultivated land, reduce biodiversity reduce available water for humans and wildlife, harbour dangerous or destructive wildlife, and act as reservoirs of human and livestock disease.

The management of forests is often referred to as forestry. Forest management has changed considerably over the last few centuries, with rapid changes from the 1980s onwards culminating in a practice now referred to as sustainable forest management. Forest ecologists concentrate on forest patterns and processes, usually with the aim of elucidating cause-and-effect relationships. Foresters who practice sustainable forest management focus on the integration of ecological, social, and economic values, often in consultation with local communities and other stakeholders.

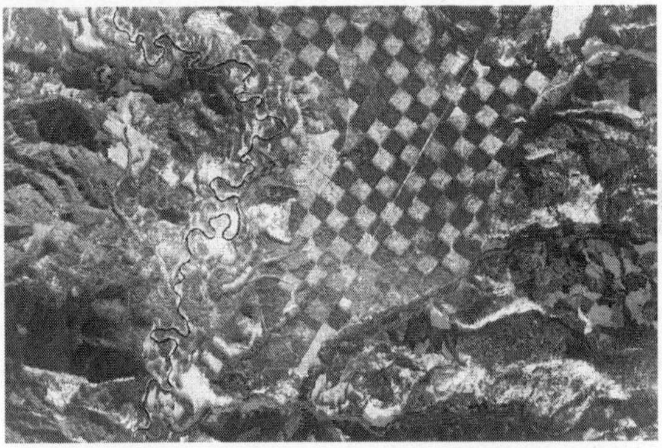

Figure: Priest River winding through Whitetail Butte with lots of forestry to the east—these lot patterns have existed since the mid-19th century. The white patches reflect areas with younger, smaller trees, where winter snow cover shows up brightly to the astronauts. Dark green-brown squares are parcels

Humans have generally decreased the amount of forest worldwide. Anthropogenic factors that can affect forests include logging, urban sprawl, human-caused forest fires, acid rain, invasive species, and the slash and burn practices of swidden agriculture or shifting cultivation. The loss and re-growth of forest leads to a distinction between two broad types of forest, primary or old-growth forest and secondary forest. There are also many natural factors that can cause changes in forests over time including forest fires, insects, diseases, weather, competition between species, etc. In 1997, the World Resources Institute recorded that only 20% of the world's original forests remained in large intact tracts of undisturbed forest. More than 75% of these intact forests lie in three countries—the boreal forests of Russia and Canada and the rainforest of Brazil. According to FAO's *Global Forest Resources Assessment 2020*, an estimated 420 million hectares of forest has been lost worldwide through deforestation since 1990, but the rate of forest loss has declined substantially. In the most recent five-year period (2015–2020), the annual rate of deforestation was estimated at 10 million hectares, down from 12 million hectares in 2010–2015.

China instituted a ban on logging, beginning in 1998, due to the erosion and flooding that it caused. In addition, ambitious tree planting programmes in countries such as China, India, the United States and Vietnam – combined with natural expansion of forests in some regions – have added more than seven million hectares of new forests annually. As a result, the net loss of forest area was reduced to 5.2 million hectares per year between 2000 and 2010, down from 8.3 million hectares annually in the 1990s. In 2015, a study for *Nature Climate Change* showed that the trend has recently been reversed, leading to an "overall gain" in global biomass and forests. This gain is due especially to reforestation in China and Russia. However new forests are not completely equivalent to old growth forests in terms of species diversity, resilience and carbon capture. On 7 September 2015, the Food and Agriculture Organization of the United Nations released a new study stating that, over the last 25 years, the global deforestation rate has decreased by 50% due to improved management of forests and greater government protection.

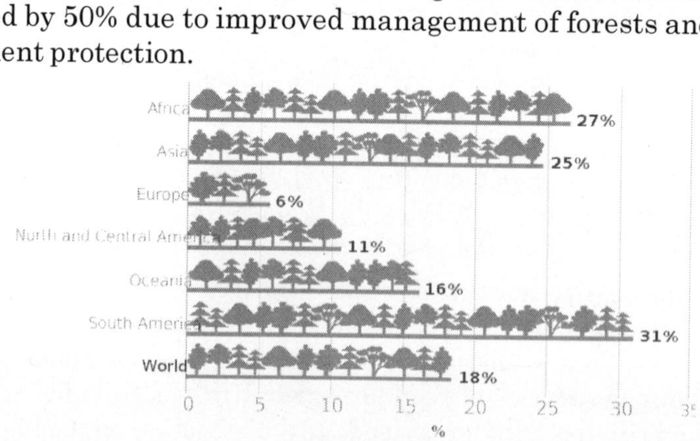

Figure : Proportion of forest in protected areas, by region, 2020

There is an estimated 726 million ha of forest in protected areas worldwide. Of the six major world regions, South America has the highest share of forests in protected areas, at 31 percent. The area of forest in protected areas globally has increased by 191 million ha since 1990, but the rate of annual increase slowed in 2010–2020.

Smaller areas of woodland in cities may be managed as urban forestry, sometimes within public parks. These are often created for human benefits; Attention Restoration Theory argues that spending time in nature reduces stress and improves health, while forest schools and kindergartens help young people to develop social as well as scientific skills in forests. These typically need to be close to where the children live, for practical logistics.

Canada

Canada has about 4,020,000 square kilometres (1,550,000 sq mi) of forest land. More than 90% of forest land is publicly owned and about 50% of the total forest area is allocated for harvesting. These allocated areas are managed using the principles of sustainable forest management, which includes extensive consultation with local stakeholders. About eight percent of Canada's forest is legally protected from resource development. Much more forest land—about 40 percent of the total forest land base—is subject to varying degrees of protection through processes such as integrated land use planning or defined management areas such as certified forests.

By December 2006, over 1,237,000 square kilometres of forest land in Canada (about half the global total) had been certified as being sustainably managed. Clearcutting, first used in the latter half of the 20th century, is less expensive, but devastating to the environment, and companies are required by law to ensure that harvested areas are adequately regenerated. Most Canadian provinces have regulations limiting the size of clear-cuts, although some older clear-cuts can range upwards of 110 square kilometres (27,000 acres) in size which was cut over several years.

Latvia

Latvia has about 3,270,000 hectares (12,626 sq mi) of forest land which equates to 50.6% of Latvia's total area (24,938 sq mi). 1,510,000 hectares of forest land (46.3% of total forest land) is publicly owned and 1,750,000 hectares of forest land (53.7% of total forest land) is in private hands. Latvia's forests have been steadily increasing over the years which is in contrast to many other nations, mostly due to the forestation of land not used for agriculture. In 1935 there was only 1,757,000 hectares of forest, today this has increased by more than 150%. Birch is the most common tree at 28.2% followed by pine (26,9%), spruce (18.3%), grey alder (9.7%), aspen (8,0%), black alder (5.7%), oak/ash (1.2%) and finally hardwood making up the rest (2.0%).

Figure: Latvian Pine Forest in Íegums Municipality

United States

In the United States, most forests have historically been affected by humans to some degree, though in recent years improved forestry practices have helped regulate or moderate large scale or severe impacts. However, the United States Forest Service estimates a net loss of about 2 million hectares (4,942,000 acres) between 1997 and 2020; this estimate includes conversion of forest land to other uses, including urban and suburban development, as well as afforestation and natural reversion of abandoned crop and pasture land to forest. However, in many areas of the United States, the area of forest is stable or increasing, particularly in many northern states. The opposite problem from flooding has plagued national forests, with loggers complaining that a lack of thinning and proper forest management has resulted in large forest fires.

Largest forests in the world

Largest forests in the world		
Forest	**Area**	**Countries**
Amazon rainforest	5,500,000 km^2 (2,100,000 sq mi)	Brazil, Peru, Colombia, Bolivia, Ecuador, French Guiana, Guyana, Suriname, Venezuela
Congo Rainforest	2,000,000 km^2 (770,000 sq mi)	Angola, Cameroon, Central African Republic, Democratic Republic of the Congo, Republic of the Congo, Equatorial Guinea, Gabon
Atlantic Forest	1,315,460 km^2 (507,900 sq mi)	Brazil, Argentina, Paraguay
Valdivian Temperate Rainforest	248,100 km^2 (95,800 sq mi)	Chile, Argentina
Tongass National Forest	68,000 km^2 (26,000 sq mi)	United States
Rainforest of Xishuangbanna	19,223 km^2 (7,422 sq mi)	China
Sunderbans	10,000 km^2 (3,900 sq mi)	India, Bangladesh
Daintree Rainforest	1,200 km^2 (460 sq mi)	Australia
Kinabalu Park	754 km^2 (291 sq mi)	Malaysia

Forests named after their creators

Some areas of land have been reforested and named after their creators, Larose Forest,and Molai forest are two so named.

Natural climate stabilizer

Forests stabilize climate and work as natural air conditioner. They work as store houses for Carbon and as per an estimate the world's forest at present stores around 289 giga tonnes of Carbon in their biomass (GFRA, 2010). Deforestation releases greenhouse gases, viz. carbon dioxide, methane and nitrous oxide, which accounts for 25 per cent of the warming effect of all greenhouse gas emissions. Replacing the carbon storage function of all tropical forests would cost an estimated $3.7 billion-equal to the gross national product of Japan. Forest, by working as natural air conditioner helps in climate stabilization.

Forests- source for recreation

Forests serve people directly for recreation. National Parks, wild life sanctuaries, bird sanctuaries and other forest associated recreation attract millions of tourists every year and earn good amount of revenue. In addition to above mentioned benefits, forests contribute significantly in the following ways-

- Forests afford protection to wildlife against strong cold (or hot) & dry winds.
- It affords protection against adverse effects of solar radiation.
- They help in balancing carbon-di-oxide & oxygen of the atmosphere.
- Forests maintain fertility of the soil by returning nutrients to it through litter.
- Forests regulate the earth's temperature regime & water cycle.
- Forests check soil erosion, landslides, shifting of sands & silting.
- Forests reduce danger of floods.
- The forests help in biological rejuvenation of soils by opening soil and improve it by adding organic litter or humus.

Forest informatics

Forest informatics is the combined science of Forestry and informatics, with a special emphasis on collection, management, and processing of data, information and knowledge, and the incorporation of informatic concepts and theories specific to enrich forest management and forest science; it has a similar relationship to library science and information science.

It is an interdisciplinary science primarily concerned with the collection, classification, manipulation, storage, retrieval and dissemination of information. Information, in this context, includes both human and machine readable documents. Examples of human readable documents include maps, field data sheets, operational schedules, and long term asset management plans with narrative text. Machine readable documents include files for geographic information systems (GIS), Global Positioning Systems (GPS), and other applications like spreadsheets, and relational database management systems.

As in management science, Forest Informatics uses decision support systems, mathematical modeling, statistics, and algorithms from engineering, operations research, computer science, and artificial intelligence to support decision-making activities. Common forestry problems include harvest scheduling, model fitting, optimal sampling, remote sensing, crew assignment, image classification, treatment timing, and log bucking problems, many of which can be formulated as optimization problems (e.g. generalized assignment problem, traveling salesman problem, knapsack problem, job shop scheduling, and vehicle routing problems). The practice includes information processing and the engineering of information systems, decision support systems, geographic information systems, and global positioning systems. The research field includes studies the structure, algorithms, behavior, and interactions of natural and artificial systems that store, process, access and communicate information about forested ecosystems.

Sustainability

Sustainability is the ability to exist constantly. In the 21st century, it refers generally to the capacity for Earth's biosphere and human civilization to co-exist. It is also defined as the process of people maintaining change in a homeostasis-balanced environment, in which the exploitation of resources, the direction of investments, the orientation of technological development, and institutional change are all in harmony and enhance both current and future potential to meet human needs and aspirations. For many in the field, sustainability is defined through the following interconnected domains or pillars: environmental, economic and social, which according to Fritjof Capra, is based on the principles of systems thinking. Sub-domains of sustainable development have been considered also: cultural, technological and political. According to *Our Common Future*, sustainable development is defined as development that "meets the needs of the present without compromising the ability of future generations to meet their own needs." Sustainable development may be the organizing principle of sustainability, yet others may view the two terms as paradoxical (seeing development as inherently unsustainable).

Sustainability can also be defined as a socio-ecological process characterized by the pursuit of a common ideal. An ideal is by definition unattainable in a given time and space. However, by persistently and dynamically approaching it, the process results in a sustainable system Many environmentalists and ecologists argue that sustainability is achieved through the balance of species and the resources within their environment. As is typically practiced in natural resource management, the goal is to maintain this equilibrium, available resources must not be depleted faster than resources are naturally generated.

Modern use of the term "sustainability" is broad and difficult to define precisely. Originally, "sustainability" meant making only such use of natural, renewable resources that people can continue to rely on their yields in the long term. The concept of sustainability, or *Nachhaltigkeit* in German, can be traced back to Hans Carl von Carlowitz (1645–1714), and was applied to forestry.

"Healthy" ecosystems and environments are necessary for the survival of humans and other organisms. Ways of reducing undesirable human impact may include environmentally-friendly chemical engineering, environmental resources management, environmental protection, and human-population control. Information is gained from green computing, green chemistry, earth science, environmental science and conservation biology. Ecological economics studies the fields of academic research that aim to address human economies and natural ecosystems.

Moving towards sustainability can involve social challenges that entail international and national law, urban planning and transport, supply-chain management, local and individual lifestyles and ethical consumerism. Ways of living more sustainably can take many forms, such as:

- reorganizing living conditions (e.g., ecovillages, eco-municipalities and sustainable cities)
- reappraising economic sectors (permaculture, green building, sustainable agriculture) or work practices (sustainable architecture)
- using science to develop new technologies (green technologies, renewable energy and sustainable fission and fusion power)
- designing systems in a flexible and reversible manner
- adjusting individual lifestyles to conserve natural resources

In sum, "the term 'sustainability' should be viewed as humanity's target goal of human-ecosystem equilibrium (homeostasis), while 'sustainable development' refers to the holistic approach and temporal processes that lead us to the endpoint of sustainability". Despite the increased popularity of the use of the term "sustainability", the possibility that human societies will achieve environmental sustainability has been, and continues to be, questioned—in light of environmental degradation, climate change, overconsumption, population growth and societies' pursuit of unlimited economic growth in a closed system.

Sustainable forest management

Sustainable forest management (SFM) is the management of forests according to the principles of sustainable development. Sustainable forest management has to keep the balance between three main pillars: ecological, economic and socio-cultural. Successfully achieving sustainable forest management will provide integrated benefits to all, ranging from safeguarding local livelihoods to protecting biodiversity and ecosystems provided by forests, reducing rural poverty and mitigating some of the effects of climate change. Forest conservation is essential to stop climate change.

Feeding humanity and conserving and sustainably using ecosystems are complementary and closely interdependent goals. Forests supply water, mitigate climate change and provide habitats for many pollinators, which are essential for sustainable food production. It is estimated that 75 percent of the world's leading food crops, representing 35 percent of global food production, benefit from animal pollination for fruit, vegetable or seed production.

The "Forest Principles" adopted at the Earth Summit (United Nations Conference on Environment and Development) in Rio de Janeiro in 1992 captured the general international understanding of sustainable forest management at that time. A number of sets of criteria and indicators have since been developed to evaluate the achievement of SFM at the global, regional, country and management unit level. These were all attempts to codify and provide for assessment of the degree to which the broader objectives of sustainable forest management are being achieved in practice. In 2007, the United Nations General Assembly adopted the Non-Legally Binding Instrument on All Types of Forests. The instrument was the first of its kind, and reflected the strong international commitment to promote implementation of sustainable forest management through a new approach that brings all stakeholders together.

The Sustainable Development Goal 15 is also a global initiative aimed at promoting the implementation of sustainable forest management.

Van Vigyan Kendra

Van Vigyan Kendra (VVK) or **Forest Science Centres (FSC)** has been established by Indian Council of Forestry Research and Education (ICFRE) of the Ministry of Environment and Forests Govt. of India. It intends to help disseminate various technologies developed by farmers, forest based industries and forest research institutes.

Functions

The **Van Vigyan Kendra** caters to the needs of field research of silviculture, tree improvement, soil & water conservation techniques, and afforestation techniques for saline land, techniques for forestry extension, organic farming and composting techniques, sustainable land-use systems and introduction and evaluation of both timber and non-timber species.

There will be emphasis on developing superior planting material to enhance the productivity of Seedlings, which will aid the people associated with forestry and agriculture.

A number of trainings were provided on various themes including Awareness programme on Forestry Research and its utilization and protection of forests.

CHAPTER 2

THE DIMINISHING FOREST COVER AND FOREST FIRE

Introduction

Forest cover in general refers to the relative (in percent) or sure (in square kilometres/square miles) land area that is covered by forests.

According to the Food and Agriculture Organization, a forest is defined as land spanning more than 0.5 hectares with **trees** higher than 5 meters and a **canopy cover** of more than 10 percent, or trees able to reach these thresholds *in situ*. It does not include land that is predominantly under agricultural or urban land use.

Forest cover is one category of terrestrial land cover. Land cover is the observed physical features, both natural and manmade, that occupy the earth's immediate surface ... forest cover is defined as 25% or greater canopy closure at the Landsat pixel scale (30-m × 30-m spatial resolution) for trees >5 m in height

—Hansen et al., 2010

Global forest cover, however crucial for soil health, the water cycle, climate and air quality it is, is severely threatened by deforestation, as a direct consequence of agriculture, grazing, and mining. Forest cover can be increased by reforestation and afforestation efforts, but it is very virtually impossible to restore the full range of ecological services once natural forests are converted to other land uses.

Since the onset of agriculture (about 12,000 years ago), the number of trees worldwide has dropped by 46%, according to one research published in 2018.

Global forest cover now has been estimated to be just 31% or 40 million square kilometres (4.3×10^{14} sq ft) in 2006 with 12-yearly losses (2000-2012) amounting to 2.3 million square kilometres (2.5×10^{13} sq ft)

and reforestation gains about 0.8 million square kilometres (8.6×10^{12} sq ft). According to FAO's *Global Forest Resources Assessment 2020*, the world has a total forest area of 4.06 billion hectares (ha), which is 31 percent of the total land area.

More than half of the world's forests are found in only five countries (Brazil, Canada, China, Russian Federation and United States of America).

The largest part of the forest (45 percent) is found in the tropical domain, followed by the boreal, temperate and subtropical domains. These domains are further divided into terrestrial global ecological zones, 20 of which contain some forest cover. Almost half the forest area (49 percent) is relatively intact, while 9 percent is found in fragments with little or no connectivity. Tropical rainforests and boreal coniferous forests are the least fragmented, whereas subtropical dry forest and temperate oceanic forests are among the most fragmented. Roughly 80 percent of the world's forest area is found in patches larger than 1 million hectares. The remaining 20 percent is located in more than 34 million patches across the world – the vast majority less than 1 000 hectares in size.

The world's total growing stock of trees decreased slightly, from 560 billion m3 in 1990 to 557 billion m3 in 2020, due to a net decrease in forest area. On the other hand, growing stock is increasing per unit area globally and in all regions; it rose from 132 m3 per ha in 1990 to 137 m3 per ha in 2020. Growing stock per unit area is highest in the tropical forests of South and Central America and West and Central Africa.

More than half (54%) of the world's forests is in only five countries – the Russian Federation, Brazil, Canada, the United States of America and China.

Once, half of the earth's landmass was covered with forests. Today, this cover has been reduced to one-third. A map of the world's forest cover, produced by the World Conservation Monitoring Centre and World Wide Fund for Nature (WWF), shows that 94% of it has no protection. Human intervention is now reducing that cover by 1% per year.

Forests have had historical similarities throughout the world. When the population was small, forests were plentiful and provided most of the basic needs. With the shift from a nomadic existence to a settled livelihood by man, forests were cleared for agriculture and habitations.

In due course of time, society recognized the need to manage forests as a resource for its benefit and earning revenue, ultimately resulting into forest depletion. When forests were plentiful, the impact of deforestation was not much and the eco-balance between human being and its environment was still maintained. However, with increasing need and demand for forest produce this impact became more visible and deteriorating. Though deforestation / forest degradation has been a worldwide phenomenon, it has affected developing countries more adversely. The poor socio-economic condition of the people in these countries (increasing their dependency on natural forests) has been the major cause for deterioration.

Global Scenario

Before pre-agriculture period, forests in the world were almost intact. With the invasion of agriculture, the prehistoric man started clearing forest for farming land leading to non-ending process of deforestation, which persists till date. As per an estimate, since pre-agriculture period almost 900 million hectares of forestland has been deforested by man. The historical evidences suggest that situation was not much deteriorated till 1000 years ago, when the forests were almost stable and covered around 34 per cent of the land. But with the onset of industrialization and demand for land and timber, forest cover began growing threadbare in scattered parts of Europe, Central America, China and India. Two centuries ago, the thinning spread, leaving parts of Europe and China bare, while only a century ago, and in the wake of the Industrial Revolution, eastern North America was deforested. Still little had changed and forests covered almost 32 per cent of the land during the period. However, in the recent past i.e. 50 years ago this change has accelerated explosively. Vast tracts of forest have vanished from the Far East and the mainland of South-East Asia; Central America, western North America and eastern South America, the Indian subcontinent and sub-Saharan Africa, the Amazon Basin and Central Europe. As per the Food and Agriculture Organization-U.N. "Production Year Book", 1987, during the period 1975 to 1985, the average loss of the forest has been about two percent. During these ten years, about eighty million hectares of the forest was lost. According to the GFRA Report, the situation improved a bit after 1990. The global forest area was reduced by around 8.3 million hectares (0.2%) per year between 1990 and 2000 and by around 5.2 million hectares (0.1%) per year between 2000 and 2010. The forest area has reduced in most regions since 1990, except Europe (where the area increased in both the decades) and Asia (where the area reduced between 1990 and 2000 but has increased between 2000 and 2010) (GFRA, 2010). Between 2000 and 2010, the largest decrease in forest area was in Brazil (2.6 million hectares per year on average) and the largest increase was in China (3.0 million hectares per year on average).

The overall impact of large scale deforestation has been very devastating, while on one hand the indigenous people in the tropical forests have faced poverty and alienation due to diminishing supplies of forest products and farm yields, on the other; it caused immense loss to environment and ecology. There have been profound ecological effects of forest loss, as evident in the exacerbation of droughts and floods, release of heat trapping temperatures, advent of new pests into cropped lands, much sedimentation in river beds and hydroelectric reservoirs, and loss of productive fisheries. The decline in forests along with other adverse effects also threatened the genetic diversity of the world's plants and animals. The World Conservation Union calculated that about 12.5 per cent of the world's 270,000 species of plants and about 75 per cent of the world's mammals are threatened by forest decline. The Commission concluded that "forests can no longer be used in the same way as they have been in the past. Forest products and services must be assured through new political choices and policy decisions that ensure the survival of forests."

Growing pressure on forest in India

Due to various monogenetic reasons, the pressure on natural forest is increasing throughout the world. The forests today have at least five times more pressure than what they can withstand. Like many other developing countries, forests in India are also under tremendous pressure mainly as a result of increased demand for forest produce viz. fuel, fodder, timber, non timber forest produce (NTFP) etc. by ever exploiting population of the country. Presently a major chunk of forest area in the country is under illegal encroachment. The collection of fuel wood is considerably higher than what can sustainably be removed from the forests. Forests also contribute major percentage of the fodder requirement for the cattle, including 178 million tons of green fodder and 145 million tons of dry fodder. It has been estimated that half of the livestock population i.e. 270 millions graze in forests. This demand gets accentuated due to the extraction of green fodder to the tune of 175-200 million ton annually. Currently people occupy around 1.5 million hectares of forestland for agricultural purposes. The trend of diversion of forestlands for non-forestry purposes though arrested after the Forest Conservation Act, 1980 came into force, still continues. These effects cumulatively cause loss of ecological stability and bio-diversity, reduction in carbon sink capability, climate change, floods, droughts, desertification, damages to watersheds, silting of reservoirs, estuarine & river beds, changes in hydrological regimes, etc. According to World Bank, approximately 10 million hectares of land mass is reported to be under shifting cultivation, involving several million people spread over 16 states in India. According to Forest Survey of India estimate 53-54% of the forest area is annually affected by forest fires incidence (mostly due to practices such as agriculture and shifting cultivation). The maximum forest loss in India has been estimated between 1950 and 1980, just before the enactment of Forest Conservation Act. During this period, a huge forest area was allotted to various sectors in the name of development. The non-forestry uses, for which forest area was converted during the period included agriculture, river valley projects, industries/ townships, transmission, roads etc.

A rough picture based on available information from different reports is that the gross annual depletion rate during 1980-90 was 339,000 hectares or a total of 3.39 million hectares for the ten years period. During the same period, there were afforestation efforts too, which added about one million hectare annually to the plantation area. While comparing various assessments made in Forest Survey of India Report, it is visible that as a result of various efforts made at government and other levels after 1997, the forest cover in India showed a positive sign. Though there may be some technical reasons also for this growth in the data, yet the forest cover in the country in general shows a positive change.

Despite increase in the forest cover, the overall forest ecosystems in India are reeling under acute form of degradation. The direct causes of this degradation are poverty, landlessness, derivation of livelihood from forests, lack of land use planning, uncertainties in land tenure system, biotic interference, inadequate institutional capacity, lack of restrictive covenants and punitive legislation and last but not the least the frequent forest fire incidences. While efforts are being made to increase forest area through tree planting activities, loss of natural forests continues to be the main concern for the country. Tree planting and social forestry activities cannot fully compensate for the loss of natural forests, since they have very different attributes. The remaining natural forests, and as well as the forest plantations) are becoming progressively degraded due to overuse and lack of protection. The deterioration of the quality, stocking condition and productivity of forests is a matter of serious concern, as is deforestation.

Forest degradation and its impact

Despite that the overall forest cover in the country is on increase, most of the forest areas in the country are ecologically in various stages of retrogression. The forest ecosystems are reeling under acute form of degradation, which has adversely affected the Indian society, both socially and economically. Owing to various other factors, the deterioration of the forest is the major cause for increase in both physical as well as socio-economic vulnerability of country to disasters. It has been widely accepted that deforestation increases the intensity of natural disasters and is often the factor that transforms a natural hazard or climatic extreme into a disaster.

In ecologically more sensitive and destabilized areas like Himalayas and Western Ghats, the impact of deforestation/ degradation has been more severe. In the recent past, as a result of so called developmental activities, particularly in sensitive regions, the environment has been very adversely affected, resulting into exponential increase in fragility of land mass. Deforestation and other allied land degradation activities such as water logging, flooding, ravines, shifting cultivation, mining, salinisation, soil erosion, landslides / rock falls and desertification have affected more than half of the total geographical area of the country.

The deforestation in the form of mangrove removal has made coastal regions of the country susceptible to erosion and damage to human settlements. The deforestation and destruction of other natural resources have compelled people from rural areas to migrate in search of new livelihood in urban areas, resulting into unplanned settlement even in high risk zones i.e. seismic zones, steep hillsides and flood-prone areas. The mushroom growths of slums in various metropolitan cities of the country are also at maximum risk to natural and human induced disasters. In ecologically fragile and more sensitive or destabilized areas like Himalayas, the impact of deforestation has been more severe, which directly or indirectly affect the lower plains of the country. Deforestation in Himalayas has increased the severity of floods during rainy season and reduced stream flows and dried up springs during dry seasons. The increase of soil erosion has reduced water carrying capacity of the rivers resulting into shallowing of riverbeds leading to floods in the plains. Ever increasing population pressure and increasing demand for food and fodder, with no alternative source of livelihood have forced the people to convert forest land, even in very sensitive zones, to agricultural fields, causing enormous loss to environment and stability of the region. Rapid deforestation in the Himalayas and resulting degradation of its ecology has posed a potential threat to the greenery of the Indo-Gangetic belt, causing sporadic floods in one and drought in other area. About two-third of the agricultural land of the country is affected by drought and about one-fifth is very frequently exposed to floods. Large scale industrialization, deforestation and non-sustainable overuse of other natural resources have increased pollution leading to global change in the environment. This human induced climate change has increased the overall global temperature, resulting into excessive melting of glaciers in the Himalayas and consequent increase in water-related disasters. The main reason for this climate change or increase in temperature is excessive burning of fuels (coal, petrol, diesel etc.), large-scale deforestation and forest fires, resulting into substantial increase of carbon dioxide and other harmful gases in the atmosphere. This increased carbon dioxide prevents heat rays returning back to the atmosphere. Thus a part of the heat energy absorbed by the earth and its objects partially remains in the atmosphere, resulting into overall increase of global temperature. This phenomenon called "Green House Effect" along with excessive snow melting causes many other adverse impacts on environment and society. The global warming, up to an extent is responsible for the rainfall variation, accelerating desertification and land degradation. Loss in agriculture production (particularly in rain-fed regions), human and cattle health deterioration, scarcity of water resources are other adverse impact of this phenomenon on society.

Deforestation

Figure: Satellite image of deforestation in progress in eastern Bolivia. Worldwide, 10% of wilderness areas were lost between ***1990*** *and* ***2015****.*

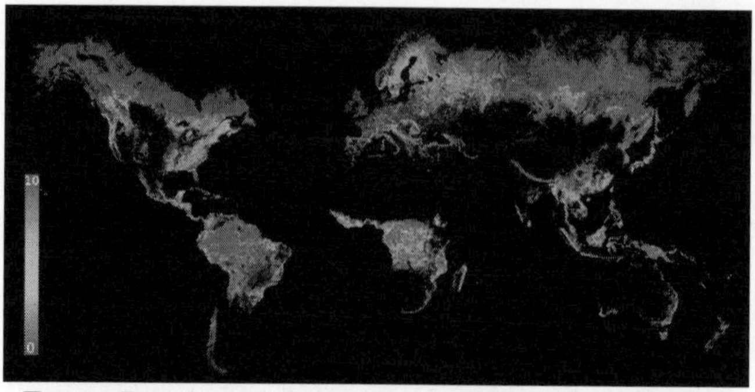

Figure: Forest Landscape Integrity Index showing anthropogenic modification of remaining forest.

Figure: Dry seasons, exacerbated by climate change, and the use of slash-and-burn methods for clearing tropical forest for agriculture, livestock, logging, and mining are causing wildfires, deforestating the Amazon rainforest.

Deforestation, **clearance**, **clearcutting,** or **clearing** is the removal of a forest or stand of trees from land that is then converted to non-forest use. Deforestation can involve conversion of forest land to farms, ranches, or urban use. The most concentrated deforestation occurs in tropical rainforests. About 31% of Earth's land surface is covered by forests. Between 15 million to 18 million hectares of forest, an area the size of Belgium, are destroyed every year, on average 2,400 trees are cut down each minute.

The Food and Agriculture Organization of the United Nations defines deforestation as the conversion of forest to other land uses (regardless of whether it is human-induced). "Deforestation" and "forest area net change" are not the same: the latter is the sum of all forest losses (deforestation) and all forest gains (forest expansion) in a given period. Net change, therefore, can be positive or negative, depending on whether gains exceed losses, or vice versa.

For instance, FAO estimate that the global forest carbon stock has decreased 0.9%, and tree cover 4.2% between 1990 and 2020. The forest carbon stock in Europe (including Russia) increased from 158.7 to 172.4 Gt between 1990 and 2020. In North America, the forest carbon stock increased from 136.6 to 140 Gt in the same period. However, carbon stock decreased from 94.3 to 80.9 Gt in Africa, 45.8 to 41.5 Gt in South and Southeast Asia combined, 33.4 to 33.1 Gt in Oceania, 5 to 4.1 Gt in Central America, and from 161.8 to 144.8 Gt in South America. The IPCC (Intergovernmental Panel on Climate Change) states that there is disagreement about whether the global forest is shrinking or not, and quote research indicating that tree cover has increased 7.1% between 1982 and 2016.[a] IPCC also writes: «While above-ground biomass carbon stocks are estimated to be declining in the tropics, they are increasing globally due to increasing stocks in temperate and boreal forests [...].»[10]

Agricultural expansion continues to be the main driver of deforestation and forest fragmentation and the associated loss of forest biodiversity. Large-scale commercial agriculture (primarily cattle ranching and cultivation of soya bean and oil palm) accounted for 40 percent of tropical deforestation between 2000 and 2010, and local subsistence agriculture for another 33 percent. Trees are cut down for use as building material, timber or sold as fuel (sometimes in the form of charcoal or timber), while cleared land is used as pasture for livestock and agricultural crops. The vast majority of agricultural activity resulting in deforestation is subsidized by government tax revenue. Disregard of ascribed value, lax forest management, and deficient environmental laws are some of the factors that lead to large-scale deforestation. Deforestation in many countries—both naturally occurring and human-induced—is an ongoing issue. Between 2000 and 2012, 2.3 million square kilometres (890,000 sq mi) of forests around the world were cut down. Deforestation and forest degradation continue to take place at alarming rates, which contributes significantly to the ongoing loss of biodiversity.

The removal of trees without sufficient reforestation has resulted in habitat damage, biodiversity loss, and aridity. Deforestation causes extinction, changes to climatic conditions, desertification, and displacement of populations, as observed by current conditions and in the past through the fossil record. Deforestation also has adverse impacts on biosequestration of atmospheric carbon dioxide, increasing negative feedback cycles contributing to global warming. Global warming also puts increased pressure on communities who seek food security by clearing forests for agricultural use and reducing arable land more generally. Deforested regions typically incur significant other environmental effects such as adverse soil erosion and degradation into wasteland.

The resilience of human food systems and their capacity to adapt to future change depends on that very biodiversity – including dryland-adapted shrub and tree species that help combat desertification, forest-dwelling insects, bats and bird species that pollinate crops, trees with extensive root systems in mountain ecosystems that prevent soil erosion, and mangrove species that provide resilience against flooding in coastal areas. With climate change exacerbating the risks to food systems, the role of forests in capturing and storing carbon and mitigating climate change is of ever-increasing importance for the agricultural sector.

According to a study published in Scientific Reports if deforestation continue in current rate in the next 20 – 40 years, it can trigger a full or almost full extinction of humanity. To avoid it humanity should pass from a civilization dominated by the economy to "cultural society" that "privileges the interest of the ecosystem above the individual interest of its components, but eventually in accordance with the overall communal interest"

Deforestation is more extreme in tropical and subtropical forests in emerging economies. More than half of all plant and land animal species in the world live in tropical forests. As a result of deforestation, only 6.2 million square kilometres (2.4 million square miles) remain of the original 16 million square kilometres (6 million square miles) of tropical rainforest that formerly covered the Earth. An area the size of a football pitch is cleared from the Amazon rainforest every minute, with 136 million acres (55 million hectares) of rainforest cleared for animal agriculture overall. More than 3.6 million hectares of virgin tropical forest was lost in 2018. Consumption and production of beef is the primary driver of deforestation in the Amazon, with around 80% of all converted land being used to rear cattle. 91% of Amazon land deforested since 1970 has been converted to cattle ranching. The global annual net loss of trees is estimated to be approximately 10 billion. According to the Global Forest Resources Assessment 2020 the global average annual deforested land in the 2015–2020 demi-decade was 10 million hectares and the average annual forest area net loss in the 2000–2010 decade was 4.7 million hectares. The world has lost 178 million ha of forest since 1990, which is an area about the size of Libya.

Causes

According to the United Nations Framework Convention on Climate Change (UNFCCC) secretariat, the overwhelming direct cause of deforestation is agriculture. Subsistence farming is responsible for 48% of deforestation; commercial agriculture is responsible for 32%; logging is responsible for 14%, and fuel wood removals make up 5%.

Experts do not agree on whether industrial logging is an important contributor to global deforestation. Some argue that poor people are more likely to clear forest because they have no alternatives, others that the poor lack the ability to pay for the materials and labour needed to clear forest. One study found that population increases due to high fertility rates were a primary driver of tropical deforestation in only 8% of cases.

Other causes of contemporary deforestation may include corruption of government institutions, the inequitable distribution of wealth and power, population growth and overpopulation, and urbanization. Globalization is often viewed as another root cause of deforestation, though there are cases in which the impacts of globalization (new flows of labor, capital, commodities, and ideas) have promoted localized forest recovery.

Another cause of deforestation is climate change. 23% of tree cover losses result from wildfires and climate change increase their frequency and power. The rising temperatures cause massive wildfires especially in the Boreal forests. One possible effect is the change of the forest composition.

Figure: Deforestation in the Maranhão state of Brazil, 2016

In 2000 the United Nations Food and Agriculture Organization (FAO) found that "the role of population dynamics in a local setting may vary from decisive to negligible", and that deforestation can result from "a combination of population pressure and stagnating economic, social and technological conditions".

The degradation of forest ecosystems has also been traced to economic incentives that make forest conversion appear more profitable than forest conservation. Many important forest functions have no markets, and hence, no economic value that is readily apparent to the forests' owners or the communities that rely on forests for their well-being. From the perspective of the developing world, the benefits of forest as carbon sinks or biodiversity reserves go primarily to richer developed nations and there is insufficient compensation for these services. Developing countries feel that some countries in the developed world, such as the United States of America, cut down their forests centuries ago and benefited economically from this deforestation, and that it is hypocritical to deny developing countries the same opportunities, i.e. that the poor should not have to bear the cost of preservation when the rich created the problem.

Some commentators have noted a shift in the drivers of deforestation over the past 30 years. Whereas deforestation was primarily driven by subsistence activities and government-sponsored development projects like transmigration in countries like Indonesia and colonization in Latin America, India, Java, and so on, during the late 19th century and the earlier half of the 20th century, by the 1990s the majority of deforestation was caused by industrial factors, including extractive industries, large-scale cattle ranching, and extensive agriculture. Since 2001, commodity-driven deforestation, which is more likely to be permanent, has accounted for about a quarter of all forest disturbance, and this loss has been concentrated in South America and Southeast Asia.

Causes of deforestation/ forest degradation

Due to various factors, mostly anthropogenic, Indian forests have been experiencing significant loss since long. In the colonial period the forests were exploited mainly for timber. During Second World War, a major portion of good quality timber was exported from India for making warships and for certain other purposes. The introduction of Railways in India was also to some extent responsible for large-scale forest cutting, mainly for railway sleepers. Along with the increasing population pressure, the low productivity and imbalance between demand and supply are the other major factors for forest loss in the country. The forests of India have a growing stock of 4, 740 million m3 with an annual increment of 87.62 million m3. Since 78% of the forest area is subjected to grazing, as well as heavy removal of forest products and 51% is subjected to occasional fires (the net annual loss being about 74,000 hectares), the productivity of the forests is rather low. The paradox is that forest produce 70% timber and 30% of fuel wood, while the demand for wood is around 70% as fuel wood and 30% for timber; further exacerbating the situation. Despite its richness in variety and species, if compared at a global scale, the forest cover in the country is very less. Human and cattle population explosion around forest land and increase in demand of various forest produce have adversely affected the Indian forests.

Agriculture (including shifting cultivation), multipurpose projects, irrigation structures, industries, mining, quarrying for minerals, road construction, erection of transmission lines, and clearing for encroachments (industrialization, urbanization and human settlements) etc. have been some of the major factors responsible for the present poor situation of forests in the country. In addition to these direct and visible reasons, many other socio-political and socio- economic factors responsible for poor state of the forests of the country are- poverty and unemployment, poor legislative provisions, impractical management tactics, gender inequality etc. Unlike the developed countries, the forest in the country is inextricably linked with socio-economic status of the society. The people living within or near the forests are directly or indirectly dependent on forests for their daily needs. Till recent past, due to low population pressure, this relationship had been very harmonious, however, with increasing demand, the situation deteriorated and as a result forests suffered adversely.

Demographic pressure and poor socio-economic condition

As mentioned earlier too, the increasing demographic pressure has been the major factor for deforestation. The growth in population has increased the demand for livelihood. Scarcity of other livelihood resources, especially in rural areas has forced people to exploit natural resources, particularly the forests. Along with demographic pressure, poverty has been another important factor, responsible for forest degradation. As per the Planning Commission of India about 37% of the country population is forced to survive below poverty line. Poor in the country are compelled to exploit natural resources for their survival. Dependency on forests for fuel, wood, grazing, encroachment etc. is the issues that are directly linked with the poverty.

Increasing demand for fuel wood and timber

Fuel wood burning is the major source of energy for cooking and other small scale industries in India .The consumption of fuel wood in our country is about five times higher than what can sustainably be removed from forests. Fuel wood meets about 40% energy needs of the country. The estimated fuel wood consumption in the country is about 320 million tones. About 70% of the fuel wood is accounted for by households and the rest by commercial and industrial units. Around 80% of the rural people and some 48% of urban people use fuel wood for cooking.

As per Forest Survey of India Report, during 1996 the total consumption of fuel wood in the country was 201 million tons, out of which 103 million tons was directly extracted from the forest and rest 98 million tons was made available from farm forestry sector. Out of 103 million tons extracted from the forest, only 17 million tons is available on sustainable basis and the rest 86 million tons is removed in excess, causing enormous damage to forest and its eco system. The demand and supply data for fuel wood indicates an increasing gap. As major portion for this gap is filled from over exploitation of forest, beyond its carrying capacity, it gives an alarming signal for forest health. Not only fuel wood, the timber extraction is other major problem associated with forest degradation in the country. According to estimates made, about 64 million cubic meters timber was consumed in the country during 1996. Out of this, around fifty percent was directly extracted from the forests and rest was from the farm forestry.

Cattle grazing and fodder supply

India has the largest cattle population in the world. In the absence of adequate productive pasturelands and appropriate grazing policy, forests have become the major source of grazing and fodder. Grazing not only directly harm the forest, it also increase soil erosion and deteriorate soil quality, thus having adverse impact on forest regeneration and growth. As per an estimate, around 60% of the livestock (about 300 million) graze in forests. These include traditional sedentary village livestock and migratory animals herded by ethnic grazers. Additionally, grazers collect about 175 million tones of green fodder annually, by lopping and harvesting grasses, which adversely affect regeneration of forests. A sample survey by Forest Survey of India estimates that the grazing affects approximately 78 per cent of country's forests. Grazing occurs even in protected areas (67 percent of national parks and 83% of wildlife sanctuaries surveyed reported grazing incidences).

Illicit felling & encroachments

As a result of increased demand of wood, the pressure on forest has gone up, encouraging illicit felling. Illicit felling by timber mafia, fuel wood gatherers, land encroachers etc. is very common in India. The increasing demand of wood products and its increasing value has made illicit felling a profitable business for timber mafia. The sandalwood, teak, rosewood, sals etc. are the major species, which are smuggled from forest by organised gangs. Due to increasing profit involved, these gangs are usually armed with sophisticated arms and ammunitions, which make the work of forest protection very difficult and risky. In some states, these gangs have become so powerful and daring that they do not hesitate in attacking senior forest and police officials also. In addition to illicit felling, forest land encroachment for agriculture and other purposes has geared up in the recent past. The escalating land price has further accelerated this practice in rural and urban areas both. The most evident impact of weak forest protection practices is encroachment by violating forest boundaries through organized massive efforts. Boundaries also recede through illegal peripheral occupation by communities living around the forests. Though exact figures are not available for the cumulative deforested area resulting from forest encroachments, it is estimated that people illegally occupy about 1.5 million hectares of forest area for agriculture and other uses.

Development vs. Deforestation

With increasing population and subsequent development/ industrialization, the demand of land has increased substantially. The major impact of this has been on forestland. One of the major factors leading to deforestation and / or forest degradation is transfer of forest land for various non-forestry uses. This forest land diversion, particularly for developmental projects has been very harmful for forest ecosystem. A development project starts destroying the forest even before it goes into production. The process begins with building of infrastructure in the form of roads, railway lines, offices, township, electricity etc. Construction of roads in fragile belt of hilly areas affects the stability of hill slopes; damage the vegetation cover and the environment. The construction of roads has adversely affected the eco-balance in all the hill states of Himalayas and also in some southern States of Nilgiri hills.

After China War in 1962, the main emphasis of the Central Government was to increase connectivity in Himalayan region through roads. During this period even the interior villages were connected by roads. But the most unfortunate part of this development had been that these roads were constructed without giving much consideration to environmental issues. This activity has generated huge amount of debris and other waste, which ultimately became major factor for forest degradation and other environment related problems.

Industrialization and urbanization are two other major causes for forest degradation. After independence, availability of natural resource (forest, minerals, raw material etc.), electricity, communication and cheap labor have encouraged prospective industrialists to establish industries in various towns and cities of the country, resulting into environmental problems like pollution, deforestation etc.

Dam construction, obstructing the natural water flow, not only affects the ecological balance in an area but also causes environmental loss to natural resources and wildlife. Adversely attacking the forest regeneration and growth, construction of dams causes massive damage to forest and wildlife. A large part of the forest gets drowned under the reservoir. The labour force also causes destruction to forest for its fuel need. There is change in water table in the rocks/soil along the reservoir leading to increased soil erosion. The other reason for increased rate of erosion is clearance of vegetation in the upstream side. Thus all unscientific development practices result into enormous loss to vegetation and environment. Keeping in view the severity of the problem the Government of India in 1980 enacted the Forest Conservation Act (FCA) with an objective to check diversion of forestland for non-forestry purposes. Forest Conservation Act stipulates that State governments or any other authority is prevented from diverting forest lands for non-forestry purposes and that prior approval of the central government is required for any such diversion. Wherever, diversion of forest land is unavoidable for developmental purposes, approvals granted to concern State governments are subject to safeguards, which include compensatory afforestation on non-forest land with equivalent characteristics or double the area of forest land cleared. As result of enactment of Forest Conservation Act the rate of land diversion, approximately 1, 50,000 hectares per year between 1950 and 1980, fell to 25,000 hectares per year between 1980 and 1995. There has been a further decrease in the area of forest land diverted for non-forestry uses in the last few years to around 15,500 hectares per year.

Along with developmental projects, industrialization has been other major reason for large scale deforestation/ forest degradation in the country. Despite modernization and shift of dependency from wood to steel and plastic, wood is being used for several industrial processes, such as making boxes, crates, packing cases, furniture, match boxes, paper, pulp, plywood etc. The paper industry accounts for about 2% of country's total annual consumption of wood.

Shifting Cultivation

Shifting cultivation or Jhum farming, characterized by rotation of fields rather than crops, is a thousand of years old practice. Also known as slash and burn method of farming, this practice is responsible for about 5,000 km2 of deforestation annually. According to archaeological evidence, the practice of shifting cultivation can be traced as far back as 7000 BC. Shifting cultivation has been practiced throughout the world covering many tropical countries in Africa, South America, Oceania and Southeast Asia. Different estimates for the area involved under-shifting cultivation in India ranges from 5 million hectares to 11.5 million hectares. Though there is no consensus on the number of people involved in shifting cultivation, estimate ranges from 3 to 26 million. In India, shifting cultivation is practiced at least in 16 States, predominantly in the northeastern parts in the country. The increase in the number of shifting cultivators and the declining productivity of forest soils have resulted in the shifting cultivation cycle shortening from about 20 years to 4 years. Such intense farming on poor quality forest soils has made such region ecologically very unstable. The other serious issue with the shifting cultivation tradition is that till recent past restricted to few sections of society, it is being adopted by more and more people. Many rural people, without secure access to land are also becoming shifting cultivators. Shifting cultivation is the major cause of forest fire. According to Forest Survey of India estimate (1995), 53 to 54 per cent of the forest area in India is annually affected by the incidence of forest fires (mostly from agricultural practices such as shifting cultivation).

Forest fire- the major cause of forest degradation

Along with other factors discussed, forest fire is a major cause of injury and loss to forests. With the population increase, the frequency and subsequent damage due to forest fire is increasing day by day. The impact of the fire is diverse on the forest ecosystem. Besides directly damaging the forest trees, the fire also adversely affects forest regeneration, microclimate, soil erosion, and wild life etc. In most of the cases, the forest fire causes retrogression of forest vegetation. Forest fire is one of the major degenerating factors, which extensively damages the growing stock and its generations and making area vulnerable to erosion. It has wide-ranging adverse ecological, economic and social implications. Globally speaking, forest fires all over the world are under reported due to various factors. As per the information compiled in GFRA- 2010, on an average one percent of all forests were reported to be significantly affected by forest fire each year. However, the areas affected by fires are severely underreported, with information missing from many countries. Forest fires in India are generally ground fires. About 35 million hectares of forest area is affected by fires annually. About 95 per cent of the forest fires are caused by human beings, especially to promote new flush of grasses, collection of minor forest produce or to prepare land for shifting cultivation. While statistical data on fire loss in India are very weak, it is estimated that the proportion of the forest areas prone to forest fire annually ranges from 33% to over 90% in different states. As per an estimate of the United Nations Development Programme and Food and Agriculture Organisation project in Maharashtra State, the economic loss due to forest fire is around Rs. 9000/- per hectare per annum. If multiplied to the total forest cover of the country, this comes to a very substantial amount. The severity of the problem may be judged from the forest fire data of the year 1995 and 1999 in the two States- Uttar Pradesh and Himachal Pradesh where forest worth crores of rupees was turned to ashes during the period.

CHAPTER 3

FOREST FIRE AND ITS IMPACTS

Introduction

A forest fire, **wildfire**, **bushfire**, **wildland fire** or **rural fire** is an unplanned, unwanted, uncontrolled fire in an area of combustible vegetation starting in rural areas and urban areas. Depending on the type of vegetation present, a wildfire can also be classified more specifically as a forest fire, brush fire, bushfire (in Australia), desert fire, grass fire, hill fire, peat fire, prairie fire, vegetation fire, or veld fire. Many organizations consider **wildfire** to mean an unplanned and unwanted fire, while **wildland-fire** is a broader term that includes prescribed fire as well as wildland fire use (WFU; these are also called monitored response fires).

Figure : The Rim Fire burned more than 250,000 acres (1,000 km^2) of forest near Yosemite National Park, in 2013

Fossil charcoal indicates that wildfires began soon after the appearance of terrestrial plants 420 million years ago. The occurrence of wildfires throughout the history of terrestrial life invites conjecture that fire must have had pronounced evolutionary effects on most ecosystems' flora and fauna. Earth is an intrinsically flammable planet owing to its cover of carbon-rich vegetation, seasonally dry climates, atmospheric oxygen, and widespread lightning and volcanic ignitions.

Wildfires can be characterized in terms of the cause of ignition, their physical properties, the combustible material present, and the effect of weather on the fire. Wildfires can cause damage to property and human life, although naturally occurring wildfires may have beneficial effects on native vegetation, animals, and ecosystems that have evolved with fire. Wildfire behavior and severity result from a combination of factors such as available fuels, physical setting, and weather. Analyses of historical meteorological data and national fire records in western North America show the primacy of climate in driving large regional fires via wet periods that create substantial fuels, or drought and warming that extend conducive fire weather.

High-severity wildfire creates complex early seral forest habitat (also called "snag forest habitat"), which often has higher species richness and diversity than an unburned old forest. Many plant species depend on the effects of fire for growth and reproduction. Wildfires in ecosystems where wildfire is uncommon or where non-native vegetation has encroached may have strongly negative ecological effects.

Wildfires are among the most common forms of natural disaster in some regions, including Siberia, California, and Australia. Areas with Mediterranean climates or in the taiga biome are particularly susceptible.

In the United States and other countries, aggressive wildfire suppression aimed at minimizing fire has contributed to accumulation of fuel loads, increasing the risk of large, catastrophic fires. In the United States especially, this wildfire suppression curtailed traditional land management methods practiced by Indigenous Peoples. Modern forest management taking an ecological perspective engages in controlled burns to mitigate this risk and promote natural forest life cycles.

Causes

Natural

Leading natural causes of wildfires include:

- dry climate

- lightning
- volcanic eruption

Human activity

The most common direct human causes of wildfire ignition include arson, discarded cigarettes, power-lines arcs (as detected by arc mapping), and sparks from equipment. Ignition of wildland fires via contact with hot rifle-bullet fragments is also possible under the right conditions. Wildfires can also be started in communities experiencing shifting cultivation, where land is cleared quickly and farmed until the soil loses fertility, and slash and burn clearing. Forested areas cleared by logging encourage the dominance of flammable grasses, and abandoned logging roads overgrown by vegetation may act as fire corridors. Annual grassland fires in southern Vietnam stem in part from the destruction of forested areas by US military herbicides, explosives, and mechanical land-clearing and -burning operations during the Vietnam War.

Prevalence

The most common cause of wildfires varies throughout the world. In Canada and northwest China, lightning operates as the major source of ignition. In other parts of the world, human involvement is a major contributor. In Africa, Central America, Fiji, Mexico, New Zealand, South America, and Southeast Asia, wildfires can be attributed to human activities such as agriculture, animal husbandry, and land-conversion burning. In China and in the Mediterranean Basin, human carelessness is a major cause of wildfires. In the United States and Australia, the source of wildfires can be traced both to lightning strikes and to human activities (such as machinery sparks, cast-away cigarette butts, or arson). Coal seam fires burn in the thousands around the world, such as those in Burning Mountain, New South Wales; Centralia, Pennsylvania; and several coal-sustained fires in China. They can also flare up unexpectedly and ignite nearby flammable material.

Human-caused wildfires account for 40% of wildfires in British Columbia, and are caused by activities such as open burning, the use of engines or vehicles, dropping burning substances such as cigarettes, or any other human-related activities that can create a spark or a heat source sufficient to ignite a wildfire. Hundreds of fires were burning in 2019 in British Columbia and a quarter of them were caused by humans.

Spread

Figure: A surface fire in the western desert of Utah, United States

The spread of wildfires varies based on the flammable material present, its vertical arrangement and moisture content, and weather conditions. Fuel arrangement and density is governed in part by topography, as land shape determines factors such as available sunlight and water for plant growth. Overall, fire types can be generally characterized by their fuels as follows:

- **Ground** fires are fed by subterranean roots, duff and other buried organic matter. This fuel type is especially susceptible to ignition due to spotting. Ground fires typically burn by smoldering, and can burn slowly for days to months, such as peat fires in Kalimantan and Eastern Sumatra, Indonesia, which resulted from a riceland creation project that unintentionally drained and dried the peat.
- **Crawling** or **surface** fires are fueled by low-lying vegetation on the forest floor such as leaf and timber litter, debris, grass, and low-lying shrubbery. This kind of fire often burns at a relatively lower temperature than crown fires (less than 400 °C (752 °F)) and may spread at slow rate, though steep slopes and wind can accelerate the rate of spread.
- **Ladder** fires consume material between low-level vegetation

and tree canopies, such as small trees, downed logs, and vines. Kudzu, Old World climbing fern, and other invasive plants that scale trees may also encourage ladder fires.

- **Crown**, **canopy**, or **aerial** fires burn suspended material at the canopy level, such as tall trees, vines, and mosses. The ignition of a crown fire, termed *crowning*, is dependent on the density of the suspended material, canopy height, canopy continuity, sufficient surface and ladder fires, vegetation moisture content, and weather conditions during the blaze. Stand-replacing fires lit by humans can spread into the Amazon rain forest, damaging ecosystems not particularly suited for heat or arid conditions.

In monsoonal areas of north Australia, surface fires can spread, including across intended firebreaks, by burning or smoldering pieces of wood or burning tufts of grass carried intentionally by large flying birds accustomed to catch prey flushed out by wildfires. Species implicated are Black Kite (*Milvus migrans*), Whistling Kite (*Haliastur sphenurus*), and Brown Falcon (*Falco berigora*). Local Aborigines have known of this behavior for a long time, including in their mythology.

Physical properties

Figure: Experimental fire in Canada

Figure: A dirt road acted as a fire barrier in South Africa. The effects of the barrier can clearly be seen on the unburnt (left) and burnt (right) sides of the road.

Wildfires occur when all the necessary elements of a fire triforce come together in a susceptible area: an ignition source is brought into contact with a combustible material such as vegetation, that is subjected to enough heat and has an adequate supply of oxygen from the ambient air. A high moisture content usually prevents ignition and slows propagation, because higher temperatures are needed to evaporate any water in the material and heat the material to its fire point. Dense forests usually provide more shade, resulting in lower ambient temperatures and greater humidity, and are therefore less susceptible to wildfires. Less dense material such as grasses and leaves are easier to ignite because they contain less water than denser material such as branches and trunks. Plants continuously lose water by evapotranspiration, but water loss is usually balanced by water absorbed from the soil, humidity, or rain. When this balance is not maintained, plants dry out and are therefore more flammable, often a consequence of droughts.

A wildfire *front* is the portion sustaining continuous flaming combustion, where unburned material meets active flames, or the smoldering transition between unburned and burned material. As the front approaches, the fire heats both the surrounding air and woody material through convection and thermal radiation. First, wood is dried as water is vaporized at a temperature of 100 °C (212 °F). Next, the pyrolysis of wood at 230 °C (450 °F) releases flammable gases. Finally, wood can smolder at 380 °C (720 °F) or, when heated sufficiently, ignite at 590 °C (1,000 °F). Even before the flames of a wildfire arrive at a particular location, heat transfer from the wildfire front warms the air to 800 °C (1,470 °F), which pre-heats and dries flammable materials, causing materials to ignite faster and allowing the fire to spread faster. High-temperature and long-duration surface wildfires may encourage flashover or *torching*: the drying of tree canopies and their subsequent ignition from below.

Wildfires have a rapid *forward rate of spread* (FROS) when burning through dense uninterrupted fuels. They can move as fast as 10.8 kilometres per hour (6.7 mph) in forests and 22 kilometres per hour (14 mph) in grasslands. Wildfires can advance tangential to the main front to form a *flanking* front, or burn in the opposite direction of the main front by *backing*. They may also spread by *jumping* or *spotting* as winds and vertical convection columns carry *firebrands* (hot wood embers) and other burning materials through the air over roads, rivers, and other barriers that may otherwise act as firebreaks. Torching and fires in tree canopies encourage spotting, and dry ground fuels around a wildfire are especially vulnerable to ignition from firebrands. Spotting can create *spot fires* as hot embers and firebrands ignite fuels downwind from the fire. In Australian bushfires, spot fires are known to occur as far as 20 kilometres (12 mi) from the fire front.

The incidence of large, uncontained wildfires in North America has increased in recent years, significantly impacting both urban and agriculturally-focused areas. The physical damage and health pressures left in the wake of uncontrolled fires has especially devastated farm and ranch operators in affected areas, prompting concern from the community of healthcare providers and advocates servicing this specialized occupational population.

Especially large wildfires may affect air currents in their immediate vicinities by the stack effect: air rises as it is heated, and large wildfires create powerful updrafts that will draw in new, cooler air from surrounding areas in thermal columns. Great vertical differences in temperature and humidity encourage pyrocumulus clouds, strong winds, and fire whirls with the force of tornadoes at speeds of more than 80 kilometres per hour (50 mph). Rapid rates of spread, prolific crowning or spotting, the presence of fire whirls, and strong convection columns signify extreme conditions. The thermal heat from a wildfire can cause significant weathering of rocks and boulders, heat can rapidly expand a boulder and thermal shock can occur, which may cause an object's structure to fail.

Forest fire through ages

Fire has been a major influencing factor on the development and management of many of the world's forests. Some forest ecosystems have evolved in response to frequent fires from natural causes, but most others are susceptible to the effects of wildfire. Each year millions of hectares of the world's forests are consumed by fire, which results in enormous economic losses because of burnt timber; degraded real estate; high costs of suppression; damage to environmental, recreational and amnesty values; and loss of life. Forest fires are not new to the world; they have burned across the earth for millions of years. Evidence of fires that burned in the past exists today in petrified trees that lived long ago and have over many years turned into hard rock. Some petrified trees have fossilized charcoal called "Fusain" in their trunks. The charcoal marks, which indicate that the tree was once in the path of a fire, are called fire scars on a living tree. All these evidences support that our temperate world's forest ecosystem has developed with fire.

During prehistoric period, mankind used fire as a cultural tool. Of the many ways in which pre-historic man modified the forest, fire was effective enough to significantly influence the forest types. Fire, today is the single most important tool in converting tropical forests to agricultural land. Man in prehistoric time used fire to clear fields for cultivation, to drive game, and as a means of communication. Thus forest fire since prehistoric time has been a part of man's cultural traditions and was not of much damage to forest ecosystem. Actually small and limited forest fires are integral part of the forest ecosystem and are very essential for healthy and proper growth and development of forests. However, the large uncontrolled fire badly damages the forest in many ways because of its increased intensity and frequency and also its uncontrollable nature. Forest fire has really become a matter of much concern for society.

Forest fire – part of ecosystem

Fires in forests are not unnatural. It has been a natural part of the ecosystem since origin of forest on this planet. Most of the fires are very useful and essential for good natural forest development and regeneration. Throughout historic time forest fires have been ignited and burned naturally through the forest. These low intensity fires in past kept the forest floor free from the natural annual build up of the litter i.e. tree needles, dead grass, senescent leaves & twigs, thick brush, and dead trees. As a result, fire has shaped vegetation patterns and wildlife distributions in the forests.

Fire effects on all forests are not equal. While same fire, beneficial for one ecosystem, may be dreadful for the other, depending upon the climatic conditions and type of vegetation. Tropical rain forests choked in fog and continuously drenched by mists and down pours are least affected by fires. In deciduous forests of temperate region, as a result of heavy rainfall and dampness and relatively high humidity, fire is very occasional and less damaging. Evergreen forests with broad leaves of dry areas and conifer forest are more susceptible to fire in general.

Fire severity affects the survival and establishment of many shrub species, organic matter consumption, soil heating and the biological response of soil, flora and fauna. The extent of soil heating depends on the physical characteristics of the fire, in particular to the quantity of fuel consumed. In some ecosystems, fire may be considered as a component of great importance, particularly in promoting primary productivity. Fire may also stimulate flowering, increase the branching behavior, seed production and seedling establishment. Soil heating due to fire changes its chemical, physical and microbial properties. The direct chemical changes during soil heating and combustion of soil organic matter lead to a massive volatilization of simple nitrogenous compounds, mainly nitrate and ammonium and to some extent sulphur, phosphorus, and other ions depending on the fire intensity and temperature. But at the same time, fire transforms soil nitrogen bound in organic substances into ammonium, a form readily available to either plants or subsequent microbial nitrification. The increase in ammonium and nitrate concentrations in many ecosystems has also been reported as a result of fire incidences. This increase in the availability of nutrients (nitrogen, phosphorous and other mineral nutrients) is certainly beneficial for the vegetation and enhances its growth. Prescribed burning is used as a way to put the fire back into a specific unit of land i.e. ecosystem. Fallen trees and limbs left to rot on the forest floor decay at a very slow rate. In fact, large logs can take more than hundred years to decompose. This process is aided by the numerous species of bacteria, insects, and wildlife that live in the decomposing materials. All this rotting is one way through which nature recycles nutrients back into the soil. Pine needles decompose very slowly. It takes more than a year for ten percent of the pine needles to decay. As a result, year after year, pine needles continue to build up until they are eliminated by fire. In brief, natural recycling is a very lengthy and time-consuming phenomenon in forest. Fire is the best process to intensify this natural process. Faster recycling occurs during a fire and gases are released into the atmosphere in the form of smoke.

In the burned area, nitrogen and other nutrient remains are leached back into the soil as rain soaks the ground. This is nature's way of rapidly feeding nutrients to the soil. Unfortunately, when there is too much fuel on the ground and it is burned, an intense severe fire occurs, where these benefits are often missing. Intense fire tends to scorch the ground and kill the trees. In brief, the forest needs both slow recycling from decomposition and fast recycling from fires. Thus forest fires are not always harmful. Small and controlled fires in the form of prescribed burning are very essential and useful. In the absence of fire, vegetative changes may result in fuel loads far exceeding safety levels, which would pose a serious threat to forest if ignited. The small scale controlled fires provide social and ecological benefits too, like reducing risk of catastrophic forest fire, improving silvicultural opportunities, increasing forage and habitat opportunities for wildlife, enhancing biodiversity and so on.

Fire ecology

Figure: The Old Fire burning in the San Bernardino Mountains (image taken from the International Space Station)

Fire ecology is a scientific discipline concerned with natural processes involving fire in an ecosystem and the ecological effects, the interactions between fire and the abiotic and biotic components of an ecosystem, and the role as an ecosystem process. Many ecosystems, particularly prairie, savanna, chaparral and coniferous forests, have evolved with fire as an essential contributor to habitat vitality and renewal. Many plant species in fire-affected environments require fire to germinate, establish, or to reproduce. Wildfire suppression not only eliminates these species, but also the animals that depend upon them.

Figure: A combination of photos taken at a photo point at Florida Panther NWR. The photos are panoramic and cover a ***360*** *degree view from a monitoring point. These photos range from pre-burn to 2 year post burn.*

Campaigns in the United States have historically molded public opinion to believe that wildfires are always harmful to nature. This view is based on the outdated beliefs that ecosystems progress toward an equilibrium and that any disturbance, such as fire, disrupts the harmony of nature. More recent ecological research has shown, however, that fire is an integral component in the function and biodiversity of many natural habitats, and that the organisms within these communities have adapted to withstand, and even to exploit, natural wildfire. More generally, fire is now regarded as a 'natural disturbance', similar to flooding, wind-storms, and landslides, that has driven the evolution of species and controls the characteristics of ecosystems.

Fire suppression, in combination with other human-caused environmental changes, may have resulted in unforeseen consequences for natural ecosystems. Some large wildfires in the United States have been blamed on years of fire suppression and the continuing expansion of people into fire-adapted ecosystems, but climate change is more likely responsible. Land managers are faced with tough questions regarding how to restore a natural fire regime, but allowing wildfires to burn is the least expensive and likely most effective method.

Abiotic responses

Fires can affect soils through heating and combustion processes. Depending on the temperatures of the soils caused by the combustion processes, different effects will happen- from evaporation of water at the lower temperature ranges, to the combustion of soil organic matter and formation of pyrogenic organic matter, otherwise known as charcoal.

Fires can cause changes in soil nutrients through a variety of mechanisms, which include oxidation, volatilization, erosion, and leaching by water, but the event must usually be of high temperatures in order of significant loss of nutrients to occur. However, quantity of nutrients available in soils are usually increased due to the ash that is generated, and this is made quickly available, as opposed to the slow release of nutrients by decomposition. Rock spalling (or thermal exfoliation) accelerates weathering of rock and potentially the release of some nutrients.

Increase in the pH of the soil following a fire is commonly observed, most likely due to the formation of calcium carbonate, and the subsequent decomposition of this calcium carbonate to calcium oxide when temperatures get even higher. It could also be due to the increased cation content in the soil due to the ash, which temporarily increases soil pH. Microbial activity in the soil might also increase due to the heating of soil and increased nutrient content in the soil, though studies have also found complete loss of microbes on the top layer of soil after a fire. Overall, soils become more basic (higher pH) following fires because of acid combustion. By driving novel chemical reactions at high temperatures, fire can even alter the texture and structure of soils by affecting the clay content and the soil's porosity.

Removal of vegetation following a fire can cause several effects on the soil, such as increasing the temperatures of the soil during the day due to increased solar radiation on the soil surface, and greater cooling due to loss of radiative heat at night. Fewer leaves to intercept rain will also cause more rain to reach the soil surface, and with fewer plants to absorb the water, the amount of water content in the soils might increase. However, it might be seen that ash can be water repellent when dry, and therefore water content and availability might not actually increase.

Biotic responses and adaptations

Fire adaptations are life history traits of plants and animals that help them survive wildfire or to utilize resources created by wildfire. These traits can help plants and animals increase their survival rates during a fire and/or reproduce offspring after a fire. Both plants and animals have multiple strategies for surviving and reproducing after fire.

Evolution of fire survival traits

Phylogenetic studies indicated that fire adaptive traits have evolved for a long time (tens of millions of years) and these traits are associated with the environment. In habitats with regular surface fires, similar species developed traits such as thick bark and self-pruning branches. In crown fire regimes, pines have evolved into traits such as retaining dead branches in order to attract fires. These traits are inherited from the fire-sensitive ancestors of modern pines. Other traits such as serotiny and fire-stimulating flowering also have evolved for millions of years. Some species are capable of using flammability to establish their habitats. For example, trees evolved with fire-embracing traits can "sacrifice" themselves during fires. But they also cause fires to spread and kill their less flammable neighbors. With the help of other fire adaptive traits such as serotiny, flammable trees will occupy the gap created by fires and colonize the habitat.

Plants

Plants have evolved many adaptations to cope with fire. Of these adaptations, one of the best-known is likely *pyriscence*, where maturation and release of seeds is triggered, in whole or in part, by fire or smoke; this behaviour is often erroneously called *serotiny*, although this term truly denotes the much broader category of seed release activated by any stimulus. All pyriscent plants are serotinous, but not all serotinous plants are pyriscent (some are necriscent, hygriscent, xeriscent, soliscent, or some combination thereof). On the other hand, germination of seed activated by trigger is not to be confused with pyriscence; it is known as *physiological dormancy*.

Figure: Lodgepole pine cones

In chaparral communities in Southern California, for example, some plants have leaves coated in flammable oils that encourage an intense fire. This heat causes their fire-activated seeds to germinate (an example of dormancy) and the young plants can then capitalize on the lack of competition in a burnt landscape. Other plants have smoke-activated seeds, or fire-activated buds. The cones of the Lodgepole pine (*Pinus contorta*) are, conversely, pyriscent: they are sealed with a resin that a fire melts away, releasing the seeds. Many plant species, including the shade-intolerant giant sequoia (*Sequoiadendron giganteum*), require fire to make gaps in the vegetation canopy that will let in light, allowing their seedlings to compete with the more shade-tolerant seedlings of other species, and so establish themselves. Because their stationary nature precludes any fire avoidance, plant species may only be fire-intolerant, fire-tolerant or fire-resistant.

Fire intolerance

Fire-intolerant plant species tend to be highly flammable and are destroyed completely by fire. Some of these plants and their seeds may simply fade from the community after a fire and not return; others have adapted to ensure that their offspring survives into the next generation. "Obligate seeders" are plants with large, fire-activated seed banks that germinate, grow, and mature rapidly following a fire, in order to reproduce and renew the seed bank before the next fire. Seeds may contain the receptor protein KAI2, that is activated by the growth hormones karrikin released by the fire.

Figure: Fire tolerance. Typical regrowth after an Australian bushfire

Fire tolerance

Fire-tolerant species are able to withstand a degree of burning and continue growing despite damage from fire. These plants are sometimes referred to as "resprouters." Ecologists have shown that some species of resprouters store extra energy in their roots to aid recovery and re-growth following a fire. For example, after an Australian bushfire, the Mountain Grey Gum tree (*Eucalyptus cypellocarpa*) starts producing a mass of shoots of leaves from the base of the tree all the way up the trunk towards the top, making it look like a black stick completely covered with young, green leaves.

Fire resistance

Fire-resistant plants suffer little damage during a characteristic fire regime. These include large trees whose flammable parts are high above surface fires. Mature ponderosa pine (*Pinus ponderosa*) is an example of a tree species that suffers virtually no crown damage under a naturally mild fire regime, because it sheds its lower, vulnerable branches as it matures.

Animals, birds and microbes

Figure: A mixed flock of hawks hunting in and around a bushfire

Like plants, animals display a range of abilities to cope with fire, but they differ from most plants in that they must avoid the actual fire to survive. Although birds may be vulnerable when nesting, they are generally able to escape a fire; indeed they often profit from being able to take prey fleeing from a fire and to recolonize burned areas quickly afterwards. In fact, many wildlife species globally are dependent on recurring fires in fire-dependent ecosystems to create and maintain habitat. Some anthropological and ethno-ornithological evidence suggests that certain species of fire-foraging raptors may engage in intentional fire propagation to flush out prey. Mammals are often capable of fleeing a fire, or seeking cover if they can burrow. Amphibians and reptiles may avoid flames by burrowing into the ground or using the burrows of other animals. Amphibians in particular are able to take refuge in water or very wet mud.

Some arthropods also take shelter during a fire, although the heat and smoke may actually attract some of them, to their peril. Microbial organisms in the soil vary in their heat tolerance but are more likely to be able to survive a fire the deeper they are in the soil. A low fire intensity, a quick passing of the flames and a dry soil will also help. An increase in available nutrients after the fire has passed may result in larger microbial communities than before the fire. The generally greater heat tolerance of bacteria relative to fungi makes it possible for soil microbial population diversity to change following a fire, depending on the severity of the fire, the depth of the microbes in the soil, and the presence of plant cover. Certain species of fungi, such as *Cylindrocarpon destructans* appear to be unaffected by combustion contaminants, which can inhibit re-population of burnt soil by other microorganisms, and therefore have a higher chance of surviving fire disturbance and then recolonizing and out-competing other fungal species afterwards.

Fire and ecological succession

Fire behavior is different in every ecosystem and the organisms in those ecosystems have adapted accordingly. One sweeping generality is that in all ecosystems, fire creates a mosaic of different habitat patches, with areas ranging from those having just been burned to those that have been untouched by fire for many years. This is a form of ecological succession in which a freshly burned site will progress through continuous and directional phases of colonization following the destruction caused by the fire. Ecologists usually characterize succession through the changes in vegetation that successively arise. After a fire, the first species to re-colonize will be those with seeds are already present in the soil, or those with seeds are able to travel into the burned area quickly. These are generally fast-growing herbaceous plants that require light and are intolerant of shading. As time passes, more slowly growing, shade-tolerant woody species will suppress some of the herbaceous plants. Conifers are often early successional species, while broad leaf trees frequently replace them in the absence of fire. Hence, many conifer forests are themselves dependent upon recurring fire.

Different species of plants, animals, and microbes specialize in exploiting different stages in this process of succession, and by creating these different types of patches, fire allows a greater number of species to exist within a landscape. Soil characteristics will be a factor in determining the specific nature of a fire-adapted ecosystem, as will climate and topography.

Examples of fire in different ecosystems

Forests

Mild to moderate fires burn in the forest understory, removing small trees and herbaceous groundcover. High-severity fires will burn into the crowns of the trees and kill most of the dominant vegetation. Crown fires may require support from ground fuels to maintain the fire in the forest canopy (passive crown fires), or the fire may burn in the canopy independently of any ground fuel support (an active crown fire). High-severity fire creates complex early seral forest habitat, or snag forest with high levels of biodiversity. When a forest burns frequently and thus has less plant litter build-up, below-ground soil temperatures rise only slightly and will not be lethal to roots that lie deep in the soil. Although other characteristics of a forest will influence the impact of fire upon it, factors such as climate and topography play an important role in determining fire severity and fire extent. Fires spread most widely during drought years, are most severe on upper slopes and are influenced by the type of vegetation that is growing.

Forests in British Columbia

In Canada, forests cover about 10% of the land area and yet harbor 70% of the country's bird and terrestrial mammal species. Natural fire regimes are important in maintaining a diverse assemblage of vertebrate species in up to twelve different forest types in British Columbia. Different species have adapted to exploit the different stages of succession, regrowth and habitat change that occurs following an episode of burning, such as downed trees and debris. The characteristics of the initial fire, such as its size and intensity, cause the habitat to evolve differentially afterwards and influence how vertebrate species are able to use the burned areas.

Shrublands

Figure: Lightning-sparked wildfires are frequent occurrences on shrublands and grasslands in Nevada.

Shrub fires typically concentrate in the canopy and spread continuously if the shrubs are close enough together. Shrublands are typically dry and are prone to accumulations of highly volatile fuels, especially on hillsides. Fires will follow the path of least moisture and the greatest amount of dead fuel material. Surface and below-ground soil temperatures during a burn are generally higher than those of forest fires because the centers of combustion lie closer to the ground, although this can vary greatly. Common plants in shrubland or chaparral include manzanita, chamise and Coyote Brush.

California shrublands

California shrubland, commonly known as chaparral, is a widespread plant community of low growing species, typically on arid sloping areas of the California Coast Ranges or western foothills of the Sierra Nevada. There are a number of common shrubs and tree shrub forms in this association, including salal, toyon, coffeeberry and Western poison oak. Regeneration following a fire is usually a major factor in the association of these species.

South African Fynbos shrublands

Fynbos shrublands occur in a small belt across South Africa. The plant species in this ecosystem are highly diverse, yet the majority of these species are obligate seeders, that is, a fire will cause germination of the seeds and the plants will begin a new life-cycle because of it. These plants may have coevolved into obligate seeders as a response to fire and nutrient-poor soils. Because fire is common in this ecosystem and the soil has limited nutrients, it is most efficient for plants to produce many seeds and then die in the next fire. Investing a lot of energy in roots to survive the next fire when those roots will be able to extract little extra benefit from the nutrient-poor soil would be less efficient. It is possible that the rapid generation time that these obligate seeders display has led to more rapid evolution and speciation in this ecosystem, resulting in its highly diverse plant community.

Grasslands

Grasslands burn more readily than forest and shrub ecosystems, with the fire moving through the stems and leaves of herbaceous plants and only lightly heating the underlying soil, even in cases of high intensity. In most grassland ecosystems, fire is the primary mode of decomposition, making it crucial in the recycling of nutrients. In some grassland systems, fire only became the primary mode of decomposition after the disappearance of large migratory herds of browsing or grazing megafauna driven by predator pressure. In the absence of functional communities of large migratory herds of herbivorous megafauna and attendant predators, overuse of fire to maintain grassland ecosystems may lead to excessive oxidation, loss of carbon, and desertification in susceptible climates. Some grassland ecosystems respond poorly to fire.

North American grasslands

In North America fire-adapted invasive grasses such as Bromus tectorum contribute to increased fire frequency which exerts selective pressure against native species. This is a concern for grasslands in the Western United States.

In less arid grassland presettlement fires worked in concert with grazing to create a healthy grassland ecosystem as indicated by the accumulation of soil organic matter significantly altered by fire. The tallgrass prairie ecosystem in the Flint Hills of eastern Kansas and Oklahoma is responding positively to the current use of fire in combination with grazing.

South African savanna

In the savanna of South Africa, recently burned areas have new growth that provides palatable and nutritious forage compared to older, tougher grasses. This new forage attracts large herbivores from areas of unburned and grazed grassland that has been kept short by constant grazing. On these unburned "lawns", only those plant species adapted to heavy grazing are able to persist; but the distraction provided by the newly burned areas allows grazing-intolerant grasses to grow back into the lawns that have been temporarily abandoned, so allowing these species to persist within that ecosystem.

Longleaf pine savannas

Much of the southeastern United States was once open longleaf pine forest with a rich understory of grasses, sedges, carnivorous plants and orchids. The above maps shows that these ecosystems (coded as pale blue) had the highest fire frequency of any habitat, once per decade or less. Without fire, deciduous forest trees invade, and their shade eliminates both the pines and the understory. Some of the typical plants associated with fire include Yellow Pitcher Plant and Rose pogonia. The abundance and diversity of such plants is closely related to fire frequency. Rare animals such as gopher tortoises and indigo snakes also depend upon these open grasslands and flatwoods. Hence, the restoration of fire is a priority to maintain species composition and biological diversity.

Figure : Yellow pitcher plant is dependent upon recurring fire in coastal plain savannas and flatwoods.

Fire in wetlands

Although it may seem strange, many kinds of wetlands are also influenced by fire. This usually occurs during periods of drought. In landscapes with peat soils, such as bogs, the peat substrate itself may burn, leaving holes that refill with water as new ponds. Fires that are less intense will remove accumulated litter and allow other wetland plants to regenerate from buried seeds, or from rhizomes. Wetlands that are influenced by fire include coastal marshes, wet prairies, peat bogs, floodplains, prairie marshes and flatwoods. Since wetlands can store large amounts of carbon in peat, the fire frequency of vast northern peatlands is linked to processes controlling the carbon dioxide levels of the atmosphere, and to the phenomenon of global warming. Dissolved organic carbon (DOC) is abundant in wetlands and plays a critical role in their ecology. In the Florida Everglades, a significant portion of the DOC is "dissolved charcoal" indicating that fire can play a critical role in wetland ecosystems.

Fire suppression

Fire serves many important functions within fire-adapted ecosystems. Fire plays an important role in nutrient cycling, diversity maintenance and habitat structure. The suppression of fire can lead to unforeseen changes in ecosystems that often adversely affect the plants, animals and humans that depend upon that habitat. Wildfires that deviate from a historical fire regime because of fire suppression are called "uncharacteristic fires".

Chaparral communities

In 2003, southern California witnessed powerful chaparral wildfires. Hundreds of homes and hundreds of thousands of acres of land went up in flames. Extreme fire weather (low humidity, low fuel moisture and high winds) and the accumulation of dead plant material from 8 years of drought, contributed to a catastrophic outcome. Although some have maintained that fire suppression contributed to an unnatural buildup of fuel loads, a detailed analysis of historical fire data has showed that this may not have been the case. Fire suppression activities had failed to exclude fire from the southern California chaparral. Research showing differences in fire size and frequency between southern California and Baja has been used to imply that the larger fires north of the border are the result of fire suppression, but this opinion has been challenged by numerous investigators and is no longer supported by the majority of fire ecologists.

One consequence of the fires in 2003 has been the increased density of invasive and non-native plant species that have quickly colonized burned areas, especially those that had already been burned in the previous 15 years. Because shrubs in these communities are adapted to a particular historical fire regime, altered fire regimes may change the selective pressures on plants and favor invasive and non-native species that are better able to exploit the novel post-fire conditions.

Fish impacts

The Boise National Forest is a US national forest located north and east of the city of Boise, Idaho. Following several uncharacteristically large wildfires, an immediately negative impact on fish populations was observed, posing particular danger to small and isolated fish populations. In the long term, however, fire appears to rejuvenate fish habitats by causing hydraulic changes that increase flooding and lead to silt removal and the deposition of a favorable habitat substrate. This leads to larger post-fire populations of the fish that are able to recolonize these improved areas. But although fire generally appears favorable for fish populations in these ecosystems, the more intense effects of uncharacteristic wildfires, in combination with the fragmentation of populations by human barriers to dispersal such as weirs and dams, *will* pose a threat to fish populations.

Forest fire – as management tool

Although fire has been the primary agent of deforestation, yet as a natural process it serves an important function in maintaining the health of certain ecosystems. The traditional view of fire as a destructive agent requiring immediate suppression has given way to the view that fire can and should be used to meet land management goals under specific ecological conditions. For decades, controlled burning has been used as a genuine forest management measure in the developed countries. In western countries, especially Britain, U.S.A., Canada etc. controlled fires are burnt at intervals of 10-12 years to maintain uniform growth. In South and Southeast Asia, including India, "Slash and Burn" method of farming is used by the tribals of hilly areas, in which they cut down and burn small areas of the forest and use the cleared land for cultivation. This method of burning offers them not only the cheapest means to clear the forest, but also supplies free fertilizers in the form of ash from the burnt vegetation on limited scales. Most of the limited fires are very useful and essential for good natural forest development and regeneration. Many cultures have stories about great fire creatures and recognize fire as a part of nature. The Egyptians believed story of a brightly colored bird named the Phoenix, which lived for as long as 600 years! At the end of its life the Phoenix would burn itself in a fire. The new Phoenix would then live for another 600 years.

Fire was a way for the Phoenix to renew itself. Natural resource managers use fires as a means to renew the natural environment. To protect natural resources and keep the environment healthy, managers study an area and write a fire prescription for that area. A prescription indicates when trained professionals ignite fire or how long a fire ignited by lightening will be allowed to burn. A prescription may include the information that how wet fuels must be, the maximum speed wind may be blowing or the highest outside temperature. Fire managers suggest exact fire prescriptions before burning is allowed. These fire prescriptions are based on weather, moisture content of the fuels, and how the fire can be lighted (ignition patterns).This prescribed fire called "management-ignited prescribed fires" allows natural processes to occur. Such fires are permitted usually in large parks or forests that have a prescribed fire plan (a prescription) and where humans and their property are not in danger. When a management-ignited fire occurs or when a fire is started by a lightning strike (also called a prescribed natural fire), the blaze is monitored daily by fire experts. Laws and regulations also determine when a prescribed fire may be ignited. Air quality regulations play an important role during such process. Prescribed burning takes place when laws, regulations, and forest needs are all in balance. Prescribed burnings don't take place when there are safety, health, and aesthetic concerns.

The Great Plains shortgrass prairie

A combination of heavy livestock grazing and fire-suppression has drastically altered the structure, composition, and diversity of the shortgrass prairie ecosystem on the Great Plains, allowing woody species to dominate many areas and promoting fire-intolerant invasive species. In semi-arid ecosystems where the decomposition of woody material is slow, fire is crucial for returning nutrients to the soil and allowing the grasslands to maintain their high productivity.

Although fire can occur during the growing or the dormant seasons, managed fire during the dormant season is most effective at increasing the grass and forb cover, biodiversity and plant nutrient uptake in shortgrass prairies. Managers must also take into account, however, how invasive and non-native species respond to fire if they want to restore the integrity of a native ecosystem. For example, fire can only control the invasive spotted knapweed (*Centaurea maculosa*) on the Michigan tallgrass prairie in the summer, because this is the time in the knapweed's life cycle that is most important to its reproductive growth.

Mixed conifer forests in the US Sierra Nevada

Mixed conifer forests in the United States Sierra Nevada used to have fire return intervals that ranged from 5 years up to 300 years, depending on the local climate. Lower elevations had more frequent fire return intervals, whilst higher and wetter elevations saw much longer intervals between fires. Native Americans tended to set fires during fall and winter, and land at a higher elevation was generally occupied by Native Americans only during the summer.

Finnish boreal forests

The decline of habitat area and quality has caused many species populations to be red-listed by the International Union for Conservation of Nature. According to a study on forest management of Finnish boreal forests, improving the habitat quality of areas outside reserves can help in conservation efforts of endangered deadwood-dependent beetles. These beetles and various types of fungi both need dead trees in order to survive. Old growth forests can provide this particular habitat. However, most Fennoscandian boreal forested areas are used for timber and therefore are unprotected. The use of controlled burning and tree retention of a forested area with deadwood was studied and its effect on the endangered beetles. The study found that after the first year of management the number of species increased in abundance and richness compared to pre-fire treatment. The abundance of beetles continued to increase the following year in sites where tree retention was high and deadwood was abundant. The correlation between forest fire management and increased beetle populations shows a key to conserving these red-listed species.

Australian eucalypt forests

Much of the old growth eucalypt forest in Australia is designated for conservation. Management of these forests is important because species like *Eucalyptus grandis* rely on fire to survive. There are a few eucalypt species that do not have a lignotuber, a root swelling structure that contains buds where new shoots can then sprout. During a fire a lignotuber is helpful in the reestablishment of the plant. Because some eucalypts do not have this particular mechanism, forest fire management can be helpful by creating rich soil, killing competitors, and allowing seeds to be released.

Forest fire- a bad master

"Fire is a good servant but a bad master" the saying is true for forest fire too. Limited and controlled forest fires have been very useful and essential for healthy forest growth. But uncontrolled forest fire may engulf and destroy healthy thick forest cover within no time. Besides direct loss to forest cover, forest fire also kills wildlife, damages environment, degrade soil quality and retrogrades forest regeneration. Since historical times, forest throughout the world has been adversely affected by fire. Fire always causes many direct or indirect effects on the forest ecosystem. They may merely be beneficial but at most of the times these effects are deteriorating. The damage to a forest by fire depends mainly on size of the fire. The main adverse impact of the uncontrollable forest fire includes damage to growing stock of forests, loss of biodiversity, increase in soil erosion, scorching of soil and reduction in its permeability and water retaining capacity and volatization of the nutrients like Nitrogen. Not only for forest vegetation and environment, the forest fire causes direct loss to human being also in the form of damage to life and property.

Extreme forest fire burns thousand of houses and kills many human beings and cattle throughout the world. As reported in the Global Forest Resources Assessment (GFRA), 2010 the recent examples of human lives loss due to forest fire include- Victoria in Australia in 2003 causing 73 fatalities and Greece fires in 2007 resulting into 70 deaths. Large uncontrolled forest fires result into health problems due to fire generated smoke. Breathing problems, skin irritation, loss of visibility and other related problems are very common during forest fires. Researchers have revealed that extreme forest fires may create conditions, which ultimately result into floods and landslides, causing enormous loss to life and property. The loss to timber increment, loss of soil fertility, soil erosion, loss of employment, drying up of water sources and loss to bio-diversity are immeasurable losses by forest fire. Forest fires during the summer of 1995 and 1999 in the Himalayan hills give an idea about the damage forest fires may result into. These fires were very severe and attracted the attention of whole nation. Fires that affected an area of 6, 77,700 hectares resulted into quantifiable timber loss worth around crores of rupees. The fires also created heavy smoke in the region, which covered the surrounding area for quite a few days. These fires caused changes in the microclimate of the area in the form of soil- moisture balance and increased evaporation. Not only the large fires, but in certain cases the small fires may also cause immense loss. In the month of February, 2001, one small fire in the Gwar Forest area of Rudraprayag district, Uttarakhand claimed four lives and injured many more. The fire broke out in a grass field, when some women were busy in harvesting the fodder grass.

Components of forest fire

Fire is the naturally occurring companion of energy released in the form of heat and light, when oxygen combines with a combustible or burnable material at a suitable high temperature (about 617 degrees F, temperature or 325 degrees C for wood to burn). There are basically three components i.e. fuel, heat and oxygen that are needed in right combination to produce fire. Combination of these components, produces the "fire triangle". By nature, triangle needs three sides, missing of any of the one side will collapse the triangle. The same is true for fire. Take away any of the three components of fire - fuel, heat or oxygen, the fire collapses. Firefighters to suppress the fire, try to do just that and remove one of the three essential components of fire.

With a steady supply of oxygen (a fire needs air that contains at least 16 percent oxygen; the earth's atmosphere contains 21 percent oxygen), fuel and temperature become critical for sustaining a fire once it is ignited. The general relationship between fuel and temperature is simple: the more fuel, higher the heat; the more heat, the faster the fire spreads. When there is plenty of heat and fuel, fires start at its own. In the words of one fire behavior expert, "Large fires live to feed themselves."Large fires can create their own winds and weather, increasing flow of oxygen. A large fire can generate hurricane – force winds with a speed of up to 120 miles an hour. The key to fire management is understanding its nature- how it is created, what it takes to create fire, and more important during difficult fire seasons- what is to be done to control it?

Fuels for forest Fire

Among the three components required for fire, the basic necessity to initiate and continue a forest fire is the presence of suitable fuel. Forest fire has different kind of fuels. The important among them are as follows:

a) Ground fuels

Ground fuel involves all the combustible material below the loose litter of the surface. The materials which constitute the ground fuels can be summarized as follows-l various decayed stages of the humus,l trees, shrubs and roots,l muck and peat.

Ground fuel always supports the glowing combustion and not the flame. They don't ignite till the moisture content drops very low (less than 20 percent). The combustion becomes very persistent once the ground fuels ignite.

b) Surface fuels

All the combustible material on the forest floor is included in surface fuels. This type of fuel is the most common type of fire fuels. This may include-

- tree leaves and the fine litter
- grasses, weeds, ferns and the other herbaceous plants,
- low brush, seedlings and saplings of trees,
- fine deadwood on the forest floor,
- large logs and stumps, andl roots of trees.

These fuels ignite very readily and provide the basic combustible material for the forest fires.

c) Aerial fuels

All the combustible dead or alive material located in the under storey and above the forest canopy is included in this type of fuels. These fuels are separated from the ground by more than a meter. The main aerial fuels include:

- branches and foliage of trees,
- trees and shrubs of the under storey,l
- standing dead trees, and
- mosses, lichens and epiphytic plants on trees.

The aerial fuels provide much needed combustible material for the spread of forest fire. The inferno in case of crown fires spreads by consuming aerial fuels.

The primary factor that helps in the spread of forest fire is the continuity of fuels. Fuel continuity is also one of the most important factors in controlling forest fire. This is due to the fact that they transfer heat by radiation, conduction and convection. Continuity, a relative term denotes both the forms of continuity i.e. vertical and horizontal. These are very essential for the fire spread and are taken into account while planning forest fire control.

Effect of climate

Figure: Lightning-sparked wildfires are frequent occurrences during the dry summer season in Nevada.

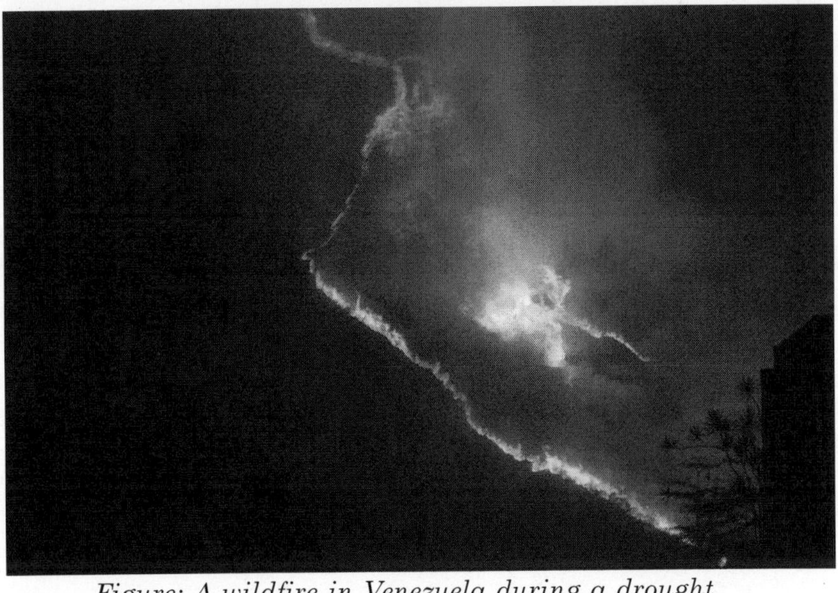

Figure: A wildfire in Venezuela during a drought

Heat waves, droughts, climate variability such as El Niño, and regional weather patterns such as high-pressure ridges can increase the risk and alter the behavior of wildfires dramatically. Years of precipitation followed by warm periods can encourage more widespread fires and longer fire seasons. Since the mid-1980s, earlier snowmelt and associated warming has also been associated with an increase in length and severity of the wildfire season, or the most fire-prone time of the year, in the Western United States. Global warming may increase the intensity and frequency of droughts in many areas, creating more intense and frequent wildfires. A 2019 study indicates that the increase in fire risk in California may be attributable to human-induced climate change. A study of alluvial sediment deposits going back over 8,000 years found warmer climate periods experienced severe droughts and stand-replacing fires and concluded climate was such a powerful influence on wildfire that trying to recreate presettlement forest structure is likely impossible in a warmer future.

Intensity also increases during daytime hours. Burn rates of smoldering logs are up to five times greater during the day due to lower humidity, increased temperatures, and increased wind speeds. Sunlight warms the ground during the day which creates air currents that travel uphill. At night the land cools, creating air currents that travel downhill. Wildfires are fanned by these winds and often follow the air currents over hills and through valleys. Fires in Europe occur frequently during the hours of 12:00 p.m. and 2:00 p.m. Wildfire suppression operations in the United States revolve around a 24-hour *fire day* that begins at 10:00 a.m. due to the predictable increase in intensity resulting from the daytime warmth.

In 2019 extreme heat and dryness caused massive wildfires in Siberia, Alaska, Canary Islands, Australia, and in the Amazon rainforest. The fires in the latter were caused mainly by illegal logging. The smoke from the fires expanded on huge territory including major cities, dramatically reducing air quality.

As of August 2020, the wildfires in the year were 13% worse than in 2019 due primarily to climate change and deforestation. The Amazon rainforest's existence is threatened by fires, some of which may be criminal arson. According to Mike Barrett, Executive Director of Science and Conservation at WWF-UK, if this rainforest is destroyed "we lose the fight against climate change. There will be no going back."

Emissions

Wildfires release large amounts of carbon dioxide, black and brown carbon particles, and ozone precursors such as volatile organic compounds and nitrogen oxides (NOx) into the atmosphere. These emissions affect radiation, clouds, and climate on regional and even global scales. Wildfires also emit substantial amounts of semi-volatile organic species that can partition from the gas phase to form secondary organic aerosol (SOA) over hours to days after emission. In addition, the formation of the other pollutants as the air is transported can lead to harmful exposures for populations in regions far away from the wildfires. While direct emissions of harmful pollutants can affect first responders and local residents, wildfire smoke can also be transported over long distances and impact air quality across local, regional, and global scales. Whether transported smoke plumes are relevant for surface air quality depends on where they exist in the atmosphere, which in turn depends on the initial injection height of the convective smoke plume into the atmosphere. Smoke that is injected above the planetary boundary layer (PBL) may be detectable from spaceborne satellites and play a role in altering the Earth's energy budget, but would not mix down to the surface where it would impact air quality and human health. Alternatively, smoke confined to a shallow PBL (through nighttime stable stratification of the atmosphere or terrain trapping) may become particularly concentrated and problematic for surface air quality. Wildfire intensity and smoke emissions are not constant throughout the fire lifetime and tend to follow a diurnal cycle that peaks in late afternoon and early evening, and which may be reasonably approximated using a monomodal or bimodal normal distribution.

Over the past century, wildfires have accounted for 20-25% of global carbon emissions, the remainder from human activities. Global carbon emissions from wildfires through August 2020 equaled the average annual emissions of the European Union. In 2020, the carbon released by California's wildfires were significantly larger than the state's other carbon emissions.[1

Environmental effects

Fires are adverse events with tangible costs for property and human life. Quantification of the immediate and direct costs of fire provide a metric for understanding the social and economic impact of fire and for assessing progress in fire prevention and protection. In addition to their physical costs, fires have a range of less immediate and obvious adverse consequences on the natural environment. These include air contamination from the fire plume (whose deposition is likely to subsequently include land and water contamination), contamination from water runoff containing toxic products, and other environmental discharges or releases from burned materials.

If an environmental event causes an impact to the ecology, generally it will affect a large area and affect the ecology of that area. This is the more likely outcome of a fire event. One can break the impacts into pathways, over which the hazards travel from the source to the target. The pathways are environmental pollution through water, air, land or noise pollution. Hazards that can be experienced from a fire include general pollutants/indicators, metals, particulates, polycyclic aromatic hydrocarbons (PAHs), chlorinate dioxins and furans, brominated dioxins and furans, polychlorinated biphenyls and polyfluorinated compounds

Atmospheric

Figure: Illegal "slash-and-burn" practice in Madagascar, 2010

Deforestation is ongoing and is shaping climate and geography.

Deforestation is a contributor to global warming, and is often cited as one of the major causes of the enhanced greenhouse effect. Tropical deforestation is responsible for approximately 20% of world greenhouse gas emissions. According to the Intergovernmental Panel on Climate Change deforestation, mainly in tropical areas, could account for up to one-third of total anthropogenic carbon dioxide emissions. But recent calculations suggest that carbon dioxide emissions from deforestation and forest degradation (excluding peatland emissions) contribute about 12% of total anthropogenic carbon dioxide emissions with a range from 6% to 17%. Deforestation causes carbon dioxide to linger in the atmosphere. As carbon dioxide accrues, it produces a layer in the atmosphere that traps radiation from the sun. The radiation converts to heat which causes global warming, which is better known as the greenhouse effect. Plants remove carbon in the form of carbon dioxide from the atmosphere during the process of photosynthesis, but release some carbon dioxide back into the atmosphere during normal respiration. Only when actively growing can a tree or forest remove carbon, by storing it in plant tissues. Both the decay and the burning of wood release much of this stored carbon back into the atmosphere. Although an accumulation of wood is generally necessary for carbon sequestration, in some forests the network of symbiotic fungi that surround the trees' roots can store a significant amount of carbon, storing it underground even if the tree which supplied it dies and decays, or is harvested and burned. Another way carbon can be sequestered by forests is for the wood to be harvested and turned into long-lived products, with new young trees replacing them. Deforestation may also cause carbon stores held in soil to be released. Forests can be either sinks or sources depending upon environmental circumstances. Mature forests alternate between being net sinks and net sources of carbon dioxide.

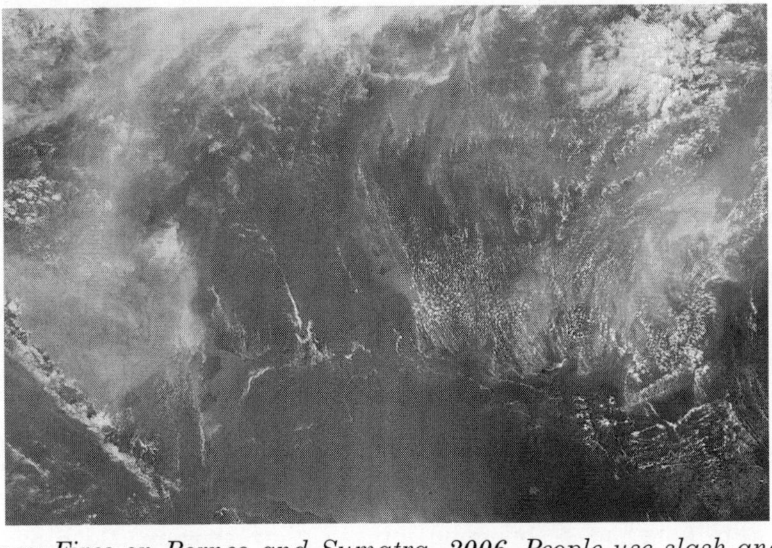

Figure: Fires on Borneo and Sumatra, 2006. People use slash-and-burn deforestation to clear land for agriculture.

In deforested areas, the land heats up faster and reaches a higher temperature, leading to localized upward motions that enhance the formation of clouds and ultimately produce more rainfall. However, according to the Geophysical Fluid Dynamics Laboratory, the models used to investigate remote responses to tropical deforestation showed a broad but mild temperature increase all through the tropical atmosphere. The model predicted <0.2 °C warming for upper air at 700 mb and 500 mb. However, the model shows no significant changes in other areas besides the Tropics. Though the model showed no significant changes to the climate in areas other than the Tropics, this may not be the case since the model has possible errors and the results are never absolutely definite. Deforestation affects wind flows, water vapour flows and absorption of solar energy thus clearly influencing local and global climate.

Reducing emissions from deforestation and forest degradation (REDD) in developing countries has emerged as a new potential to complement ongoing climate policies. The idea consists in providing financial compensations for the reduction of greenhouse gas (GHG) emissions from deforestation and forest degradation". REDD can be seen as an alternative to the emissions trading system as in the latter, polluters must pay for permits for the right to emit certain pollutants (i.e. CO2).

Rainforests are widely believed by laymen to contribute a significant amount of the world's oxygen, although it is now accepted by scientists that rainforests contribute little net oxygen to the atmosphere and deforestation has only a minor effect on atmospheric oxygen levels. However, the incineration and burning of forest plants to clear land releases large amounts of CO_2, which contributes to global warming. Scientists also state that tropical deforestation releases 1.5 billion tons of carbon each year into the atmosphere.

Hydrological

The water cycle is also affected by deforestation. Trees extract groundwater through their roots and release it into the atmosphere. When part of a forest is removed, the trees no longer transpire this water, resulting in a much drier climate. Deforestation reduces the content of water in the soil and groundwater as well as atmospheric moisture. The dry soil leads to lower water intake for the trees to extract. Deforestation reduces soil cohesion, so that erosion, flooding and landslides ensue.

Shrinking forest cover lessens the landscape's capacity to intercept, retain and transpire precipitation. Instead of trapping precipitation, which then percolates to groundwater systems, deforested areas become sources of surface water runoff, which moves much faster than subsurface flows. Forests return most of the water that falls as precipitation to the atmosphere by transpiration. In contrast, when an area is deforested, almost all precipitation is lost as run-off. That quicker transport of surface water can translate into flash flooding and more localized floods than would occur with the forest cover. Deforestation also contributes to decreased evapotranspiration, which lessens atmospheric moisture which in some cases affects precipitation levels downwind from the deforested area, as water is not recycled to downwind forests, but is lost in runoff and returns directly to the oceans. According to one study, in deforested north and northwest China, the average annual precipitation decreased by one third between the 1950s and the 1980s.

Trees, and plants in general, affect the water cycle significantly:

- their canopies intercept a proportion of precipitation, which is then evaporated back to the atmosphere (canopy interception);
- their litter, stems and trunks slow down surface runoff;
- their roots create macropores – large conduits – in the soil that increase infiltration of water;
- they contribute to terrestrial evaporation and reduce soil moisture via transpiration;
- their litter and other organic residue change soil properties that affect the capacity of soil to store water.
- their leaves control the humidity of the atmosphere by transpiring. 99% of the water absorbed by the roots moves up to the leaves and is transpired.

As a result, the presence or absence of trees can change the quantity of water on the surface, in the soil or groundwater, or in the atmosphere. This in turn changes erosion rates and the availability of water for either ecosystem functions or human services. Deforestation on lowland plains moves cloud formation and rainfall to higher elevations.

The forest may have little impact on flooding in the case of large rainfall events, which overwhelm the storage capacity of forest soil if the soils are at or close to saturation.

Tropical rainforests produce about 30% of our planet's fresh water.

Deforestation disrupts normal weather patterns creating hotter and drier weather thus increasing drought, desertification, crop failures, melting of the polar ice caps, coastal flooding and displacement of major vegetation regimes.

Soil

Due to surface plant litter, forests that are undisturbed have a minimal rate of erosion. The rate of erosion occurs from deforestation, because it decreases the amount of litter cover, which provides protection from surface runoff. The rate of erosion is around 2 metric tons per square kilometre. This can be an advantage in excessively leached tropical rain forest soils. Forestry operations themselves also increase erosion through the development of (forest) roads and the use of mechanized equipment.

Deforestation in China's Loess Plateau many years ago has led to soil erosion; this erosion has led to valleys opening up. The increase of soil in the runoff causes the Yellow River to flood and makes it yellow colored.

Greater erosion is not always a consequence of deforestation, as observed in the southwestern regions of the US. In these areas, the loss of grass due to the presence of trees and other shrubbery leads to more erosion than when trees are removed.

Soils are reinforced by the presence of trees, which secure the soil by binding their roots to soil bedrock. Due to deforestation, the removal of trees causes sloped lands to be more susceptible to landslides.

Biodiversity

Deforestation on a human scale results in decline in biodiversity, and on a natural global scale is known to cause the extinction of many species. The removal or destruction of areas of forest cover has resulted in a degraded environment with reduced biodiversity. Forests support biodiversity, providing habitat for wildlife; moreover, forests foster medicinal conservation. With forest biotopes being irreplaceable source of new drugs (such as taxol), deforestation can destroy genetic variations (such as crop resistance) irretrievably.

Since the tropical rainforests are the most diverse ecosystems on Earth and about 80% of the world's known biodiversity could be found in tropical rainforests, removal or destruction of significant areas of forest cover has resulted in a degraded environment with reduced biodiversity. A study in Rondônia, Brazil, has shown that deforestation also removes the microbial community which is involved in the recycling of nutrients, the production of clean water and the removal of pollutants.

Figure: Illegal logging in Madagascar. In 2009, the vast majority of the illegally obtained rosewood was exported to China.

It has been estimated that we are losing 137 plant, animal and insect species every single day due to rainforest deforestation, which equates to 50,000 species a year. Others state that tropical rainforest deforestation is contributing to the ongoing Holocene mass extinction. The known extinction rates from deforestation rates are very low, approximately 1 species per year from mammals and birds which extrapolates to approximately 23,000 species per year for all species. Predictions have been made that more than 40% of the animal and plant species in Southeast Asia could be wiped out in the 21st century. Such predictions were called into question by 1995 data that show that within regions of Southeast Asia much of the original forest has been converted to monospecific plantations, but that potentially endangered species are few and tree flora remains widespread and stable.

Scientific understanding of the process of extinction is insufficient to accurately make predictions about the impact of deforestation on biodiversity. Most predictions of forestry related biodiversity loss are based on species-area models, with an underlying assumption that as the forest declines species diversity will decline similarly. However, many such models have been proven to be wrong and loss of habitat does not necessarily lead to large scale loss of species. Species-area models are known to overpredict the number of species known to be threatened in areas where actual deforestation is ongoing, and greatly overpredict the number of threatened species that are widespread.

A recent study of the Brazilian Amazon predicts that despite a lack of extinctions thus far, up to 90 percent of predicted extinctions will finally occur in the next 40 years.

Health effects

Public health context

The degradation and loss of forests disrupts nature's balance. Indeed, deforestation eliminates a great number of species of plants and animals which also often results in an increase in disease, and exposure of people to zoonotic diseases. Deforestation can also create a path for non-native species to flourish such as certain types of snails, which have been correlated with an increase in schistosomiasis cases.

Forest-associated diseases include malaria, Chagas disease (also known as American trypanosomiasis), African trypanosomiasis (sleeping sickness), leishmaniasis, Lyme disease, HIV and Ebola. The majority of new infectious diseases affecting humans, including the SARS-CoV2 virus that caused the current COVID-19 pandemic, are zoonotic and their emergence may be linked to habitat loss due to forest area change and the expansion of human populations into forest areas, which both increase human exposure to wildlife.

Deforestation is occurring all over the world and has been coupled with an increase in the occurrence of disease outbreaks. In Malaysia, thousands of acres of forest have been cleared for pig farms. This has resulted in an increase in the zoonosis the Nipah virus. In Kenya, deforestation has led to an increase in malaria cases which is now the leading cause of morbidity and mortality the country. A 2017 study in the *American Economic Review* found that deforestation substantially increased the incidence of malaria in Nigeria.

Another pathway through which deforestation affects disease is the relocation and dispersion of disease-carrying hosts. This disease emergence pathway can be called "range expansion", whereby the host's range (and thereby the range of pathogens) expands to new geographic areas. Through deforestation, hosts and reservoir species are forced into neighboring habitats. Accompanying the reservoir species are pathogens that have the ability to find new hosts in previously unexposed regions. As these pathogens and species come into closer contact with humans, they are infected both directly and indirectly.

A catastrophic example of range expansion is the 1998 outbreak of Nipah virus in Malaysia. For a number of years, deforestation, drought, and subsequent fires led to a dramatic geographic shift and density of fruit bats, a reservoir for Nipah virus. Deforestation reduced the available fruiting trees in the bats' habitat, and they encroached on surrounding orchards which also happened to be the location of a large number of pigsties. The bats, through proximity spread the Nipah to pigs. While the virus infected the pigs, mortality was much lower than among humans, making the pigs a virulent host leading to the transmission of the virus to humans. This resulted in 265 reported cases of encephalitis, of which 105 resulted in death. This example provides an important lesson for the impact deforestation can have on human health.

Another example of range expansion due to deforestation and other anthropogenic habitat impacts includes the Capybara rodent in Paraguay. This rodent is the host of a number of zoonotic diseases and, while there has not yet been a human-borne outbreak due to the movement of this rodent into new regions, it offers an example of how habitat destruction through deforestation and subsequent movements of species is occurring regularly.

A now well-developed and widely-accepted theory is that the spillover of HIV from chimpanzees was at least partially due to deforestation. Rising populations created a food demand, and with deforestation opening up new areas of the forest, hunters harvested a great deal of primate bushmeat, which is believed to be the origin of HIV.

Research in Indonesia has found that outdoor workers who worked in tropical and deforested instead of tropical and naturally forested areas experienced cognitive and memory impairments which appear to be caused primarily by exposure to high heat which trees would have protected them from.

General overview

According to the World Economic Forum, 31% of emerging diseases are linked to deforestation.

According to the US Center for Disease Control and Prevention (CDC), 75% of emerging diseases in humans came from animals. The rising number of outbreaks is probably linked to habitat and biodiversity loss. In response, scientists created a new discipline, planetary health, which posits that the health of the ecosystems and the health of humans are linked. In 2015, the Rockefeller Foundation and *The Lancet* launched the concept as the Rockefeller Foundation–Lancet Commission on Planetary Health.

Since the 1980s, every decade has seen the number of new diseases in humans increase more than threefold. According to a major study by American and Australian scientists, degradation of ecosystems increases the risk of new outbreaks. The diseases that passed to humans in this way in the latest decades include HIV, Ebola, Avian flu, Swine Flu, and likely COVID-19.

In 2016, the United Nations Environment Programme published the UNEP Frontiers 2016 Report. In this report, the second chapter was dedicated to zoonotic diseases, that is diseases that pass from animals to humans. This chapter stated that deforestation, climate change, and livestock agriculture are among the main causes that increase the risk of such diseases. It mentioned that every four months, a new disease is discovered in humans. It is said that outbreaks that already happened (as of 2016) led to loss of lives and financial losses of billions dollars and if future diseases become pandemics it will cost trillions of dollars.

The report presents the causes of the emerging diseases, a large part of them environmental:

Cause	Part of emerging diseases caused by it (%)
Land-use change	31%
Agricultural industry changes	15%
International travel and commerce	13%
Medical industry changes	11%
War and Famine	7%
Climate and Weather	6%
Human demography and behavior	4%
Breakdown of public health	3%
Bushmeat	3%
Food industry change	2%
Other	4%

On page 23 of the report are presented some of the latest emerging diseases and the definite environmental cause of them:

Disease	Environmental cause
Rabies	Forest activities in South America
Bat accosiated viruses	Deforestation and Agricultural expansion
Lyme disease	Forest fragmentation in North America
Nipah virus infection	Pig farming and intensification of fruit production in Malaysia
Japanese encephalitis virus	irrigated rice production and pig farming in Southeast Asia
Ebola virus disease	Forest losses
Avian influenza	Intensive Poultry farming
SARS virus	contact with civet cats either in the wild or in live animal markets[119]

HIV/AIDS

AIDS is probably linked to deforestation. The virus firstly circulated among monkeys and apes and when the humans came and destroyed the forest and most of the primates, the virus needed a new host to survive and jumped to humans. The virus, which killed more than 25 million people, is believed to have come from the consumption of bushmeat, namely that of primates, and most likely chimpanzees in the Congo.

Malaria

Malaria, which killed 405,000 people in 2018, is probably linked to deforestation. When humans change dramatically the ecological system the diversity in mosquito species is reduced and: “”The species that survive and become dominant, for reasons that are not well understood, almost always transmit malaria better than the species that had been most abundant in the intact forests”, write Eric Chivian and Aaron Bernstein, public health experts at Harvard Medical School, in their book How Our Health Depends on Biodiversity. “This has been observed essentially everywhere malaria occurs”.

Some of the reasons for this connection, found by scientists in the latest years:

- When there is less shadow of the trees, the temperature of the water is higher which benefits mosquitos.
- When the trees don't consume water, there is more water on the ground, which also benefits mosquitos.
- Low lying vegetation is better for the species of mosquitos that transmit the disease.
- When there is no forest there is less tanin in water. Than the water is less acidic and more turbid, what is better for some species of mosquitos.
- The mosquitos that live in deforested areas are better at carrying malaria.
- Another reason is that when a large part of a forest is destroyed, the animals are crowded in the remaining fragments in higher density, which facilitate the spread of the virus between them. This leads to a bigger number of cases between animals which increase the likelihood of transmission to humans.

Consequently, the same type of mosquito bites 278 times more often in deforested areas. According to one study in Brazil, cutting of 4% of the forest, leaded to a 50% increase in Malaria cases. In one region in Peru the number of cases per year, jumped from 600 to 120,000 after people begun to cut forests.

Coronavirus disease 2019

According to the United Nations, World Health Organization and World Wildlife Foundation the Coronavirus pandemic is linked to the destruction of nature, especially to deforestation, habitat loss in general and wildlife trade.

In April 2020, United Nations Environment Programme published 2 short videos explaining the link between nature destruction, wildlife trade and the COVID-19 pandemic and created a section on its site dedicated to the issue.

The World Economic Forum published a call to involve nature recovery in the recovery efforts from the COVID-19 pandemic saying that this outbreak is linked to the destruction of the natural world.

In May 2020, a group of experts from the Intergovernmental Science-Policy Platform on Biodiversity and Ecosystem Services published an article saying that humans are the species responsible for the COVID-19 pandemic because it is linked to nature destruction and more severe epidemics might occur if humanity will not change direction. It calls to "strengthen environmental regulations; adopt a 'One Health' approach to decision-making that recognizes complex interconnections among the health of people, animals, plants, and our shared environment; and prop up health care systems in the most vulnerable countries where resources are strained and underfunded", which can prevent future epidemics and therefore is in the interest of all. The call was published on the site of the World Economic Forum.

According to the United Nations Environment Programme the Coronavirus disease 2019 is zoonotic, e.g., the virus passed from animals to humans. Such diseases are occurring more frequently in the latest decades, due to a number of factors, a large part of them environmental. One of the factors is deforestation because it reduce the space reserved for animals and destroys natural barriers between animals and humans. Another cause is climate change. Too fast changes in temperature and humidity facilitate the spread of diseases. The United Nations Environment Programme concludes that: "The most fundamental way to protect ourselves from zoonotic diseases is to prevent destruction of nature. Where ecosystems are healthy and biodiverse, they are resilient, adaptable and help to regulate diseases.

In June 2020, a scientific unit of Greenpeace with University of the West of England (UWE) published a report saying that the rise of zoonotic diseases, including coronavirus is directly linked to deforestation because it change the interaction between people and animals and reduce the amount of water necessary for hygiene and diseases treatment.

Experts say that anthropogenic deforestation, habitat loss and destruction of biodiversity may be linked to outbreaks like the COVID-19 pandemic in several ways:

- Bringing people and domestic animals in contact with a species of animals and plants that did not contacted by them before. Kate Jones, chair of ecology and biodiversity at University College London, says the disruption of pristine forests, driven by logging, mining, road building through remote places, rapid urbanisation and population growth is bringing people into closer contact with animal species they may never have been near before, resulting in transmission of new zoonotic diseases from wildlife to humans.

- Creating degraded habitats. Such habitats with a few species are more likely to cause a transmission of zoonotic viruses to humans.
- Creating more crowded habitats, with more dense population.

  Habitat loss prompts animals to search for a new one, which often results in mixing with humans and other animals.
- Disruption of ecosystems can increase the number of animals that carry many viruses, like bats and rodents. It can increase the number of mice and rats by reducing the populations of predators. Deforestation in the Amazon rainforest increase the likelihood of malaria because the deforested area is ideal for mosquitoes.
- Animal trade, by killing and transporting live and dead animals very long distances. According to American science journalist David Quammen, "We cut the trees; we kill the animals or cage them and send them to markets. We disrupt ecosystems, and we shake viruses loose from their natural hosts. When that happens, they need a new host. Often, we are it."

When climate change or deforestation causes a virus to pass to another host it became more dangerous. This is because viruses generally learn to coexist with their host and became virulent when they pass to another.

Economic impact

Figure: A satellite image showing deforestation for a palm oil plantation in Malaysia

According to the World Economic Forum, half of the global GDP is strongly or moderately dependent on nature. For every dollar spent on nature restoration, there is a profit of at least 9 dollars. Example of this link is the COVID-19 pandemic, which is linked to nature destruction and caused severe economic damage.

Damage to forests and other aspects of nature could halve living standards for the world's poor and reduce global GDP by about 7% by 2050, a report concluded at the Convention on Biological Diversity (CBD) meeting in Bonn in 2008. Historically, utilization of forest products, including timber and fuel wood, has played a key role in human societies, comparable to the roles of water and cultivable land. Today, developed countries continue to utilize timber for building houses, and wood pulp for paper. In developing countries, almost three billion people rely on wood for heating and cooking.

The forest products industry is a large part of the economy in both developed and developing countries. Short-term economic gains made by conversion of forest to agriculture, or over-exploitation of wood products, typically leads to a loss of long-term income and long-term biological productivity. West Africa, Madagascar, Southeast Asia and many other regions have experienced lower revenue because of declining timber harvests. Illegal logging causes billions of dollars of losses to national economies annually.

The new procedures to get amounts of wood are causing more harm to the economy and overpower the amount of money spent by people employed in logging. According to a study, "in most areas studied, the various ventures that prompted deforestation rarely generated more than US$5 for every ton of carbon they released and frequently returned far less than US$1". The price on the European market for an offset tied to a one-ton reduction in carbon is 23 euro (about US$35).

Rapidly growing economies also have an effect on deforestation. Most pressure will come from the world's developing countries, which have the fastest-growing populations and most rapid economic (industrial) growth. In 1995, economic growth in developing countries reached nearly 6%, compared with the 2% growth rate for developed countries. As our human population grows, new homes, communities, and expansions of cities will occur. Connecting all of the new expansions will be roads, a very important part in our daily life. Rural roads promote economic development but also facilitate deforestation. About 90% of the deforestation has occurred within 100 km of roads in most parts of the Amazon.

The European Union is one of the largest importers of products made from illegal deforestation.

Loss of life and properties-possible solutions

The economic aspect of forest fires is one of the most discussed points by the mainstream media. Vegetation, worth billions of dollars, is destroyed every year by forest fires. It is important also to consider the enormous amount of budgets the firefighting bodies are allocated each fiscal year. The measures taken are pricey for the taxpayers. Many pundits have debated the best ways to curb economic losses caused by forest fires. It is, however, important to appreciate that there are many ways of controlling financial damages caused by forest fires.

First, local authorities need to amend the already existing legislation on fireworks and other (human) causes of forest fires. Studies indicate that humans are the number one cause of forest fires, followed by natural causes, such as lighting. Second, it is crucial for the local governments to improve their response to forest fires. If the fires are handled in their early stages, huge economic losses can be avoided. In addition, the local governments should make fire reporting more efficient for the locals.

CHAPTER 4

FOREST FIRE: THE GLOBAL AND INDIAN SCENARIO

Introduction

Massive wildfires are often destructive, devastating and terrifying, which may be noted at any land around the world. With climate change, both the number and intensity of wildfires seem to have been increasing in recent years.

According to the World Meteorological Organization, the average global temperature in the summer of 2019 was 1.71 degrees Fahrenheit (0.95 degree Celsius) above the 20th century average. Many nations in Europe, Africa, Asia, America and Oceans shattered heat records. Globally, numerous alarming wildfires broke out in 2019. Some caused unprecedented damage.

Generally, most wildfires are driven by human activities and release a huge amount of greenhouse gases that contribute to climate warming. On the other hand, extreme warmth and dryness are an ideal arena for wildfires. In the era of global warming, massive wildfires are no longer local issues with risk to living environment and safety. They are also global issues that contribute to climate change. Thus, it is important for us to be aware of the scale and intensity of massive wildfires over the world in order to work out strategies to prevent and manage them.

This article presents a brief summary of major wildfires flamed over the world in 2019. Overall, wildfires have scorched an estimate of more than 37.2 million hectares (370,000 km^2) of woodland, grassland and rainforest around the world since the beginning of 2019. Here's a look at where some major fires had been burning around the world this year.

1. Australia Wildfires

Total fires: over 5,900

Burned area: Around 16.8 million hectares (42 million acres, 168000km^2)

Fatalities: 28 deaths

Australia is the sixth-largest country over the world and the largest country in Oceania, with a total land area of 7.62 million square kilometers (2.94 million square miles). According to the Forests Australia website, Australia has a total of 134 million hectares of forest grouped into 458 forest communities as of 2016, accounting for 17% of Australia's land area. Most of these forest communities are scattered in the states of New South Wales and Queensland.

Bushfires are an annual occurrence in Australia. Usually, the fire season in Australia runs from October to April. In 2019, fire season arrived early in eastern Australia. Since September 10, multiple massive bushfires began in the states of Queensland and New South Wales. Subsequently, the bushfires emerged in all the six states in Australia, destroying numerous homes. It was quite difficult to get so extensive bushfires under control. Consequently, the bushfires spread and persisted for more than 4 months. At present, many bushfires are still burning. New South Wales and Queensland are the two states most heavily affected. In New South Wales, the bushfires had burned through more land and longer period than any other blazes in recent decades, making this summer the state's worst bushfire season on record.

As of 14 January 2020, an estimated total of 16.8 million hectares (42 million acres, 168,000 square kilometers; 65,000 square miles) has been burnt across all Australian states, destroyed over 5,900 buildings and killed at least 28 people [1, 2]. A vast number of wild animals have also been killed and some endangered species may be driven to extinction. According to the European Union's Copernicus Atmosphere Monitoring Service, the 2019 bushfires in eastern Australia have already released about 400 million tons of CO2 into atmosphere – as much as two-thirds of Australia's average annual CO2 emissions or the total combined annual CO2 emissions of the 116 lowest-emitting countries .

It is speculated that climate change may be contributing to the ferocity of Australia's bushfires, since hotter and drier conditions may make the country's fire season longer and much more dangerous. Australia is one of the driest continents on Earth. Over the past two years, many areas of south-east Australia have been enduring drought due to below-average rainfall. Heatwaves associated with global warming has further exacerbated the problem. Recently, record-breaking heatwaves have hit Australia, with temperatures reaching as high as 49 °C (120 °F) in the New South Wales. In addition, the bushfires may have been spread by strong winds. Some birds in Australia can also pick up burning twigs and drop them to unburned grass to start new fires.

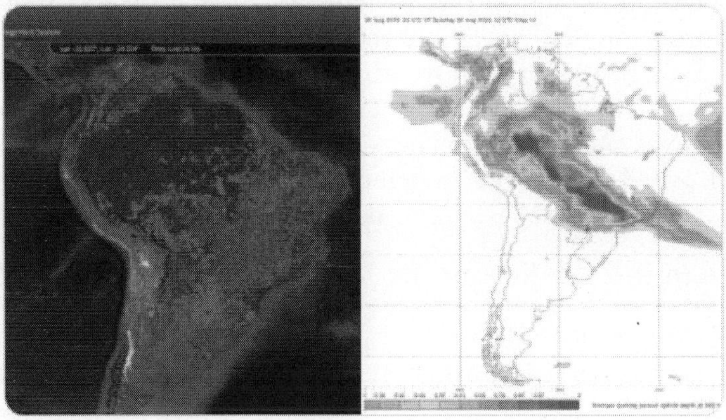

2. Amazon Wildfires

Total fires: More than 224,000

Burned area: 12 million hectares (120,000 km^2; 46,000 sq mi)

Fatalities: At least 2

Amazon rainforest is the world's largest tropical forest, covering a land area of 6.7 million km^2 (2.6 million sq mi). It's home to 30 million people and hosts 10% of the world's animals and plants. It captures up to 25% of global CO2 and generates 20% of the oxygen on Earth, making its viability a global concern.

Deforestation has been a major concern for decades in the Amazon rainforest that almost never burns on its own because of torrential rain. Each year, however, human-driven deforestation results in numerous wildfires across the Amazon area, including Brazil, Bolivia, Colombia, Peru and Venezuela, during the dry season started in May. This year, unprecedented warmth and dryness were evident across the Amazon. NOAA reported that 2019 were the second warmest year on record in the Amazon area.

The Amazon rainforest wildfires in 2019 are most alarming and terrifying. Since June 2019, an increased number of wildfires had been noted through satellite monitoring systems by the Brazil's space research center INPE. Thereafter, devastating fires were burning at a 'record rate', reported the BBC. On 11 August, Amazonas had declared a state of emergency. By August 20, the smoke plume from the fires in Rondônia and Amazonas darkened the sky above São Paulo – almost 2,800 kilometers away from the Amazon basin. Since then, wildfires in the Amazon rainforest become a major news story internationally.

By the end of August 2019, Brazil's space research center INPE reported more than 80,000 fires across all of Brazil, a 77% rise on the same period in 2018, with more than 40,000 in the Amazon area. Numerous wildfires were also recorded in surrounding nations such as Bolivia and Peru. According to the latest data reported by Euronews, more than 224,000 fires have ravaged the Amazon basin this year and 12 million hectares of forest have been burned .

3. Wildfires in the Arctic and Its Surrounding Areas in 2019

Total fires: More than 4,700

Burned area: 8.3 million hectares (83,000 km^2; 32,000 sq mi)

Fatalities: None reported

The Arctic is the area around the north pole of Earth, including Siberia (Russia), Alaska (USA), Canada and Greenland as well as the Arctic Ocean. The land within the Arctic Circle is a cold plain covered with moss and grass-like plants, while the Arctic Ocean is mostly covered with ice.

The Arctic zone may play an important role in regulating global climate. In recent years, however, the *Arctic* zone is reported to be warming faster than any other region on Earth (at a rate of almost twice the global average). It means that our Earth's resilience to global warming may be reducing.

Wildfires in the Arctic zone are common in recent years. In 2019, the Arctic is suffering its worst wildfire season on record. The World Meteorological Organization (WMO) has called the 2019 Arctic wildfires "unprecedented". Scientists estimated that more than 100 intense and long-lived wildfires in the Arctic in this year emitted 50 megatons of CO2 in June and 79 megatons of CO2 in July – more than the total released in the area in the same season between 2010 and 2018.

Wildfires in Siberia:

This year, the massive wildfires in Siberia began in June in poorly accessible northern areas. As most of the affected area is not inhabited, most of the fires are not being attended by firefighters, allowing the fires freely burning and extending for a long time. According to NASA data, the smoke from burning Siberian forests reached the territories of Alaska and Canada.

According to the Russian Aerial Forest Protection Service, quoted by RFERL, more than 5.5 million hectares (13.5 million acres; 50,000 km^2) of forests in Russia, mostly in Siberia, have burned in this summer, putting Russia on track for its worst year on record for wildfires . There have been no reported deaths or injuries due to the fires.

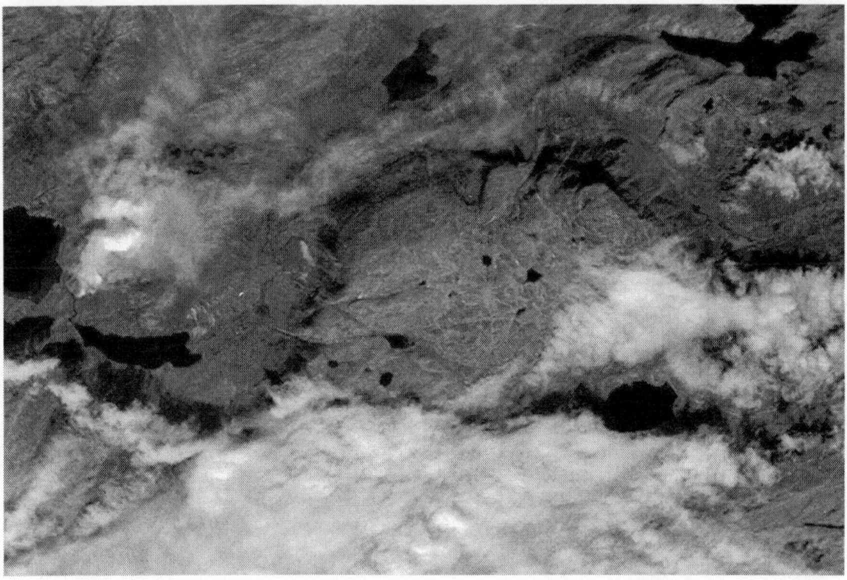

Figure: A wildfire in Greenland in July 2019. Source: NASA/CC0

Wildfires in Greenland:

In 2019, Greenland experienced an unprecedented amount and intensity of wildfires, more than in past decade combined. The largest wildfire, first detected on 10 July, took place between Sisimiut and Kangerlussuaq, Greenland. The fire persisted to August 2019. Overall, 93,000 acres (375 square kilometres) of woodland or grassland were destroyed.

Bloomberg reported that smoke from the wildfires had been billowing across the Greenland ice sheet, further driving up the already massive ice melt Greenland was experiencing amid record heat . According to the National Snow & Ice Data Center (NSIDC), "an extensive area" equivalent to about 80 billion tons of the Greenland Ice Sheet disappeared over a 10-day period in July. In total, Greenland has lost over 250 billion tons of ice in this summer .

Wildfires in Alaska:

It is said that wildfires are a normal part of life in Alaskan forests. In 2019, Alaska experienced an "extreme" fire season started from 30 April. According to Alaska Interagency Coordination Center, about 236 of the 663 fires that started this year were still burning in the state by 31 August 2019; over 2.5 million acres (1 million hectares) of forest was burned, TIME reported .

Although lightning sparks 361 of the total of 663 fires in Alaska this year, record-breaking temperatures and dried-out vegetation has also played an important role. Since May 2019, Alaska experienced "extreme" drought conditions, the US Drought Monitor reported. Record-breaking temperatures were also noted in some areas in Alaska in earlier July 2019.

*Figure: A wildfire in Canada on **16 June 2019**. Source: Government of Alberta*

Wildfires in Canada

Canada is a country northward into the Arctic Ocean. Most of its land area is dominated by forest and tundra. Many communities are nested into forests with buildings and plants intertwined, so local communities are under an increased risk of wildfires. According to

Canadian Interagency Forest Fire Centre, Canada has so far registered 3,873 wildfires in the fire season of this year. In total, these wildfires have burned more than 1.8 million hectares of forests, reported CTV News .

The largest wildfire in this year is the Chuckegg Creek blaze in northern Alberta, which burned more than 570,000 acres (2,300 km^2) of forests and led to about 10,000 people being evacuated. No fatalities were reported. Warming temperatures and increased drought are believed to be the biggest reason of the wildfires.

4. Wildfires in Europe in 2019

Total fires: More than 1,630

Burned area: Over 271,000 hectares (670,000 acres; 2,710 km^2; 1,050 sq mi)

Fatalities & Injuries: unknown.

According to Copernicus' European Forest Fire Information System (EFFIS), more than 1,626 wildfires had been recorded in the European Union this year by August 15 — more than three times the average over the past decade, the Euronews reported. More than 271,000 hectares in Europe had been ravaged by wildfires by mid-August this year.

Figure: A forest fire burning in Catalonia of Spain in June 2019

Wildfires in Spain

In this summer, Spain experienced the biggest wildfires in 20 years as heatwave griped Europe. On 26 June 2019, a wildfire broke out in the northeastern region of Catalonia. By 1st July 2019, the fire had burned across 6,500 hectares of woodland. On 29 June 2019, a wildfire broke out in Almorox, Toledo Province, and raged across 3,300 hectares (8,150 acres). From 10 to 20 August 2019, a series of forest fires broke out in the Canary Islands, resulting in the loss of large areas of the island's forests and leading to the evacuation of more than 9,000 residents from over 50 towns and villages. Overall, the wildfires across Spain destroyed at least 57,000 hectares (140, 000 acres) of forest in 2019. No injuries were reported.

Wildfires in the United Kingdom

A series of wildfires blazed in the United Kingdom from 26 February to 23 April 2019. These wildfires were remarkable since they happened at such an early date in the year. In total, the number of wildfires in the UK in 2019 was reported at 96, surpassing the previous record of 79 fires in 2018. It was estimated that over 39,537 acres (16,000 hectares) were damaged. The causes of most of the fires are attributed to much higher than average temperatures and drought conditions that have prevailed in UK since the spring of 2018.

Wildfires in Greece

From 10 August to 15 September 2019, dozens of massive wildfires broke out on different islands of Greece, making thousands of tourists and residents evacuated. A massive wildfire ripping through a "unique, untouched pine forest" on the Greek island of Evia was described as a "huge ecological disaster". At least 28,000 hectares of pine forest were destroyed, State television said. The burned area was not reported for other wildfires. According to the Greek fire department, more than 9,600 rural and urban fire calls (not all are wildfires) had been handled this year by the mid-September. No fatalities were reported. Last year, deadly wildfires killed 102 people in Greece as they spread throughout coastal Attica.

Wildfires in Portugal

It is reported that Portugal register an average of 20,000 blazes (most are not wildfires) per year between 2009 and 2018. Portuguese towns and villages are often nestled in or next to forests. As documented in a France news report , poor forest management and firefighting techniques make Portugal especially vulnerable to wildfires in 2019 as climate change makes hotter, longer summers more likely. In the mid-July 2019, three huge wildfires ravaged the mountainous Castelo Branco region of central Portugal. Strong winds and hot climate (>100 degrees Fahrenheit) drove the wildfires through villages, destroying structures and injuring 39 persons. Some 2,000 firefighters worked for several days to bring the fires under control.

5. Wildfires in California in 2019

Total fires: 7,860

Burned area: 105,140 hectares (259,800 acres; 1050 km^2)

Fatalities & Injuries: 22 injured; 5 deaths.

Description:

California is a state most vulnerable to wildfires in the United States. Each year, numerous wildfires in California are reported. In 2018, California suffered its most destructive wildfire season on record, with a total of 8,527 fires burning an area of 1.89 million acres.

In 2019, over 7,860 wildfires have been recorded in California. The largest three wildfires of this year are the Kincade Fire started on Oct. 23 (burned 77,758 acres in Sonoma county), the Walker Fire started on Sept. 4 (burned 54,518 acres in Plumas County) and the Tucker Fire began on July 28 (burned 14,150 acres in Modoc county). According to Cal Fire and the US Forest Service, a total of 105,140 hectares (259,800 acres) of forest had been burned in 2019 .

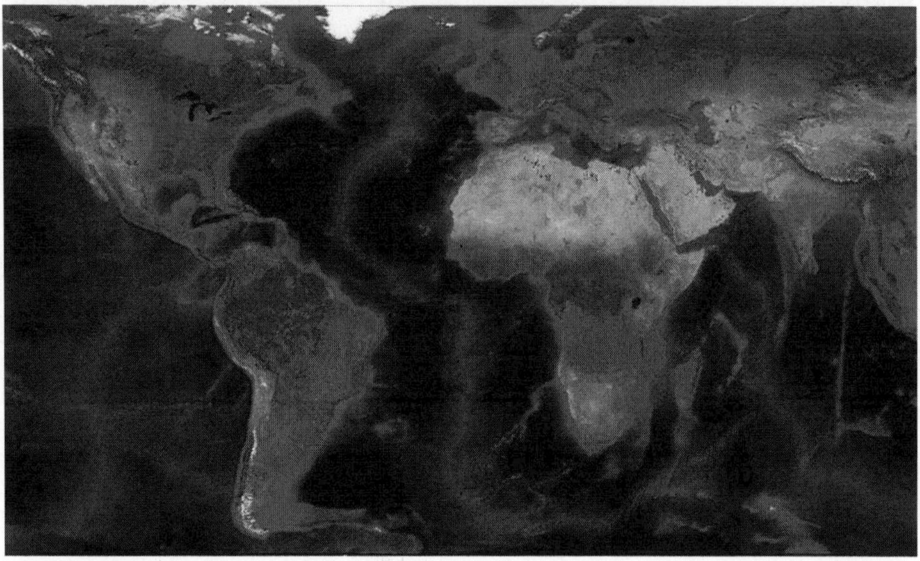

Figure: Satellite Map of the Fires in Africa on 27 August 2019

6. Wildfires in Sub-Saharan Africa

Total fires: More than 9000

Burned area: estimated > 1 million hectares

Fatalities: unknown

Sub-Saharan Africa is a region important for global climate, since the Congo Basin forest is the second largest tropical rainforest in the world. The forest is referred to as the "second green lung" of the planet after the Amazon. It covers an area of 3.3 million square kilometers in several countries including DR Congo, Gabon, Congo, Cameroon, and Central Africa.

Wildfire is quite a common event in Sub-Saharan Africa. Unlike in the Amazon where the forest burns mainly because of drought and climate change, Africa's rainforest fires are often ignited by traditional seasonal farming or agricultural activities. Farmers usually use slash-and-burn farming to clear forest in the dry season and many people use wood for cooking and energy.

According to NASA's Fire Information for Resource Management Map, there were more fires in central Africa than in Amazon this summer. Data from Copernicus, a European Union's Earth observation program, also showed that emissions from biomass burning were flagrant in the African strip across Angola, Mozambique, Zambia and the Democratic Republic of the Congo.

Although Africa is called by some researchers as the "fire continent," fire registration system is not informative or even not available in many African countries. Local African media have also given little attention to fires in their own backyard. Thus, it seems difficult to know the exact number and scale of wildfires in each African country. Using the NASA satellite imagery, TIME estimated around 6,000 fires in Angola and more than 3,000 in Congo in 2019 .

Citing a study based the satellite data from the Copernicus program, TIME reported that the total area burned in Africa in 2016 was 4.9 million square kilometers (1.89 million sq. miles) . According to the data from the Global Forest Watch, there had been more fire alerts in the Democratic Republic of Congo than in Brazil in August 2019 . So, it may be estimated that the area burned in Africa in 2019 may not be less than 1 million hectares.

7. Wildfires in Asia in 2019

Total fires: over 5100

Burned area: 814,160 acres (329,000 hectares; 3290 km^2; 1270 sq mi)

Fatalities & Injuries: at least 30 injured; 2 death.

Wildfires in Indonesia

Since July 2019, massive forest fires have ravaged Indonesia, generating thick clouds of smoke and haze that blanketed parts of Southeast Asia, sickened people and endangered wildlife. The blazes were sparked by farmers who set land on fire to make room for new crops. Washington Post reported that 185 people were arrested and accused of starting the wildfires. In total, 5,086 fires had been recorded in Indonesia this year by September 20, and more than 800,000 acres of woodland and grassland were destroyed .

Wildfires in China

According to media reports, a forest fire broke out in northwest China (Shanxi province) on 29 March 2019, destroying nearly 3,000 hectares of woodland. At least 9,000 people were evacuated. On March 30, more than 2,000 firefighters tackled a blaze in Beijing's Miyun district, which burned 50 hectares of forest. On the same day, a separate forest fire in southwest China destroyed more than 15 hectares and killed 27 firefighters and four helpers.

Wildfires in India

In February 2019, massive forest fires broke out in multiple places across a national park of the Karnataka state in India. The Indian Space Research Organization reported that the fires damaged an area of about 10,920 acres (4,420 hectares) in five days.

Wildfires in South Korea

On April 4-6, 2019, a massive wildfire broke out in Goseong County, around 210 kilometers northeast of Seoul, South Korea. The fire damaged over 200 homes and 2000 buildings across an area of 1,307 acres (5.3 km^2) and lead to two deaths, over 30 injuries and the evacuation of over 4,000 residents. It was believed that the fire was caused by a fell extra high-voltage wire. More than 13,000 firefighters were mobilized to fight the fire.

Wildfires in Vietnam

Since the mid-June, a series of wildfires have been observed in central provinces of Vietnam. More than 782 hectares of forests are damaged in three provinces. No fatalities are reported.

The year 2020 saw many massive wildfires and forest fires around the world, including the Australian bushfires which started last year and continued until March, while the US state of California was also hit by blazes that left many dead.

THE GLOBAL SCENARIO

Unusually large wildfires ravaged Alaska and Indonesia in 2015. The following year, Canada, California and Spain were devastated by uncontrolled flames. In 2017, massive fires devastated regions of Chile - and now, a deadly blaze in Portugal has claimed dozens of lives.

So, have wildfires actually increased globally, or does it just seem that way because we're tuned in more to bad news and social media?

Science suggests that over the past few decades, the number of wildfires has indeed increased, especially in the western United States. According to the Union of Concerned Scientists (UCS), every state in the western US has experienced an increase in the average annual number of large wildfires over past decades.

Extensive studies have found that large forest fires in the western US have been occurring nearly five times more often since the 1970s and 80s. Such fires are burning more than six times the land area as before, and lasting almost five times longer.

What's more, wildfire season - meaning seasons with higher wildfire potential - has universally become longer over the past 40 years.

A rapid analysis released this year found that climate change made the conditions for Australia's unprecedented 2019-20 bushfires at least 30% more likely. Further analysis – visualised below in an interactive map – has shown that, globally, climate change is driving an increase in the weather conditions that can stoke wildfires.

But despite a growing field of evidence suggesting that climate change is making the conditions for fire more likely, research finds that the total area burned by wildfires each year decreased by up to a quarter in the past two decades.

Understanding this paradox requires scientists to assess a vast range of influential factors, including climate change, human land-use and political and social motivations.

Globally, wildfires have many impacts on humans, wildlife and the economy. Wildfires are a major driver of greenhouse gas emissions and are also responsible for 5-8% of the 3.3 million annual premature deaths from poor air quality, research suggests.

Some wildfires are started naturally, chiefly by lightning. The rest are started by humans, either accidentally or by arson.

Across the world, it is estimated that just 4% of fires start naturally. However, the proportion of human-started versus lightning-started fires varies widely from region to region. For example, in the US, 84% of fires are started by humans. However, in Canada, the majority (55%) of wildfires are started by lightning.

Many regions experience distinct wildfire seasons, driven by rainy and dry periods and human practices, such as agricultural burning. However, other regions have a risk of fire year-round.

The Amazon is Still Burning

Last year, headlines all over the world were flooded with images of flames devouring the Brazilian Amazon, and the world was riveted and responsive. In 2020, obscured by the ongoing pandemic, there has been next to no coverage or public alarm about this year's fire season. But the fires in the Brazilian Amazon were up by 13% this year, making this the worst fire season in the area in a decade.

Other Amazonian countries also report a rise in both the number and severity of fires this year. At the start of October, there were a staggering 28,892 active fires active in the Brazilian Amazon alone. These fires are burning deforested areas and farmland but, most worryingly, they are increasingly affecting protected areas. In Bolivia, one of nine Amazonian countries, more than 25% of this year's major fires have burned in protected lands. This alarming trend suggests that the rainforest is getting dryer, and fast.

The Amazon is a "keystone ecosystem". That means its health and wellbeing are fundamentally tied to the health and wellbeing of our climate and planet. In short: When the forest thrives, so does our planet. When the forest suffers, our climate is at risk.

Protecting this vital ecosystem is essential because of its rich biodiversity, its large carbon stocks, its role in climate regulation, and its status as the largest remaining forest ecosystem in the world. Just as importantly, the Amazon is home to invaluable cultural diversity and to over 24 million people, many of whom belong to indigenous communities. These groups are disproportionately affected by the ongoing fires, which are burning inside their territories and threatening their survival. An increase in respiratory diseases brought on by the heavy clouds of smoke, food shortages, and an increased vulnerability to the devastating COVID-19 pandemic, have been raising great concern among Indigenous leaders.

So, what happens when the forest is on fire?

Fires fueled by deforestation caused by the expansion of the agricultural frontier and destructive policies are putting this ecosystem on the brink of an ecological tipping point. Natural systems normally change gradually; however, when there are strong pressures acting on them, a threshold emerges after which there is a sudden and usually irreversible change. Even when there is just a tiny change over the threshold, the resulting effect can be significant. These changes are driven by feedback loops that act like vicious cycles: the process driving the change feeds itself and makes the change even greater over time.

To understand this "vicious cycle" that links fires, climate, and deforestation in the Amazon, we need to know how the normal cycle works. The Amazon rainforest acts basically like a giant sponge, trapping and transporting the moisture that is blown in from the Atlantic. Once caught, each water molecule is recycled up to 6 times in the water cycle, being transported as rain and water vapor along the entire Amazon. So, in a way, the rainforest makes its own rain, transports it, and reuses it to sustain itself and to regulate the local climate.

However, when the Amazon is cut down or burned, especially around its edges in the South and East, the rainforest starts to lose its ability to trap and make rain, becoming drier and drier. This, in turn, makes it more prone to fire, and so the vicious cycle that accelerates drought and fire is complete.

Resilience... and a tipping point

The rainforest has the capacity to bounce back after changes like these. However, if these pressures become too frequent and too intense, the ecosystem is set to reach that fragile tipping point. After reaching that point, it loses its ability to recover. Once past this threshold, the forest will turn into a dry, grassy savannah that can store much less carbon.

Moreover, in the process of dying, most of the carbon that is sequestered in the forest will be released into the atmosphere, producing the equivalent of five years' worth of human carbon dioxide emissions. This, in turn, will feed the vicious cycle even more, accelerating drought and fire in the Amazon, and changing both the local and global climate.

Although a wicked problem, fires in the Amazon can be addressed through a rounded approach that focuses on cutting emissions, reducing deforestation, improving local governance, strengthening international alliances that fund the protection of the Amazon, and forging strong partnerships that recognize and apply the invaluable knowledge of the indigenous peoples who have lived and protected these rainforests for thousands of years.

Now, as we think of ways to rebuild our society after the coronavirus pandemic, we have the chance to rebuild our institutions and systems so that they focus on human and environmental well-being, on sustainability and equity and regeneration. Because, as this pandemic has shown us, human societies and the environment are inextricably linked; our health and the planet's health are one and the same.

Wildfires, forest fires around world in 2020

As countries focus on fighting the coronavirus pandemic, another crisis is unfolding around the world. This one is taking place in forests, where the latest fire seasons have been raging with unprecedented ferocity, from the Amazon to the Arctic. The number of fire outbreaks across the globe was up by 13% in April this year compared to last year, which was already a record year for fires, according to researchers. They warn that fires in 2020 are on course to be worse than in 2019.

Persistent hotter and drier weather due to climate change, and other human factors such as land conversion for agriculture and poor forest management are the main drivers behind the increase, say experts from the World Wide Fund for Nature (WWF), and the Boston Consulting Group (BCG). According to them, climate change and wildfires mutually reinforce each other, and the fires burning today in many parts of the world are bigger, more intense, and last longer than they used to. If current trends continue, there will be devastating long-term consequences on people, wildlife, and the climate, they warn. A greater number of more intense fires, for example, will release millions of extra tonnes of carbon, decimate biodiversity, destroy vital ecosystems, impact economies and people, threaten property and livelihoods, and cause severe long-term health problems for millions around the world.

The analysis comes amid wildfire reports across countries, including California where firefighters are battling some of the largest wildfires in recorded history. "The world witnessed the devastating consequences of the fires last year, from billions of wildlife lost and people losing their homes and livelihoods, not to mention the impact on climate. Proactivity and commitment must be at the heart of a global response to fires that must play out at local, sub-national, national, and regional levels. Preventing fires before they occur is paramount, and far preferable in all respects to suppressing them when they're burning. Meanwhile, good intentions on paper mean nothing if they're not followed up with real and effective actions on the ground — and these actions need to focus on forests, where the fire crisis is at its worst," writes Fran Price, global leader for forests lead at WWF.

2020 is already a year of extremes

According to the analysis, deforestation in Columbia in the first four months of 2020 suggests the area burnt could surpass the figures for all of 2019, reversing a trend of continued falls in recent decades.

In Brazil, forest loss in the first six months of the year totaled 307,000 hectares, 26% more than the same period in 2019. Increasing deforestation is likely to result in intense fires in the Amazon biome, especially in government-controlled lands, says the team. The number of fires in the Brazilian Amazon hit a 13-year high in June, at the beginning of the dry season. In July, 6,803 outbreaks of fires were detected in the Amazon, 28% more than the same period in 2019.

Figure: Deforestation in Columbia in the first four months of 2020 suggests the area burnt could surpass the figures for all of 2019, reversing a trend of continued falls in recent decades, say experts

In Bolivia, deforestation drove a 35% increase in fires from January to April 2020 compared to last year. In Australia, the 2019-2020 fire season was the worst in the country's history, and a fifth of the entire temperate broadleaf and mixed forest biome was destroyed, equivalent to an area nearly the size of England. The investigators further say that the worst wildfires for decades swept through Northern Thailand in April, destroying up to 20% of the forested area and causing 'critical levels' of air pollution in Chiang Mai. Among other incidents, the team says that the worst wildfires in the Chernobyl area's history began to burn in April, while fires across Ukraine have increased 30% this year.

The problem

Globally, fires emit carbon dioxide equivalent to the European Union every year. Their contribution to the climate crisis alone would be more than enough of a reason for the global community to treat their increase as a major threat, say experts. But that's not all: wildfires also have severe consequences for human health and wellbeing, biodiversity, and economies around the world.

The analysis reveals that wildfires affect all biomes, from forests and savannahs to grasslands and tundra. Based on records from 2000-2015, 85% of the surface area burned globally each year is located in tropical savannahs, which make up 19% of total land cover. While forests make up only 10% of the total area burned, their higher carbon storage capacity implies that they are responsible for one-quarter of all fire-related carbon dioxide emissions, explain researchers. "Forest fires in all biomes are responsible for nearly a quarter of all emissions from fires. Tropical forests are less resilient to fire, and their carbon storage contribution makes them an urgent priority for preventative action, explains the study. Fire seasons are getting longer, and extreme fire seasons are more common. From 1979 to 2013, the global fire season length increased by 19% on average. This increase is particularly severe in east Africa and Brazil, with the forests and savannahs of South America experiencing an average of over one month increase in the fire season," says the report.

Climate change is also making fire seasons increasingly unpredictable. Instances of so-called 'outliers' — abnormally long fire seasons — have increased. The word 'unprecedented' is frequently heard today when it comes to fire severity, say scientists. "California provides a striking example: over the last 100 years, five of its 20 most severe wildfires occurred between 1920-2000, another five took place from 2001-2010 — and the remaining 10 all happened in the last 10 years. In the 2019-2020 winter season, California only got half its normal levels of precipitation, while Russia has just come through its warmest winter since records began years ago, beating its 2016 record by 1.3°C," the findings suggest.

The effects of wildfires linger long after the flames die down, hitting public health and wellbeing far into the future. Every year, there are an estimated 340,000 premature deaths from respiratory and cardiovascular issues attributed to wildfire smoke. The increasing frequency and severity of wildfires pose a growing threat to biodiversity globally. Individuals, companies, and public authorities bear great economic costs due to fires, write authors. "At a time when the world is grappling with the Covid-19 pandemic, wildfire smoke increases the danger of further respiratory illnesses," they add.

The economic costs of wildfires are going up all over the world as fires increase in frequency and severity. On an individual level, fires drastically worsen living conditions, and this has economic consequences. In the US, for example, property values decrease from 10-20% in areas affected by wildfires. In 2018, the wildfire season cost the US around $50 billion in total.

The human factor

An increasing share of wildfires is due to human activity, intentional or otherwise. This is estimated to be responsible for 75% of all wildfires in recent years. The report says that in the Northern Hemisphere, most fires are caused by negligence (such as burning rubbish and debris, industrial accidents, agricultural overspill), and arson is also sometimes to blame. In Europe, negligence causes 95% of fires, and in the US, the figure is 84%.

The most at-risk locations are so-called "wildland-urban interfaces," where significant populations live in or near forests. In more remote forest areas, lightning is more likely to be responsible — in Canada's British Columbia, 40% of wildfires are traced back to human origins.

In tropical and subtropical regions, forest fires are mostly intentionally set for land-use change, clearing and preparing new areas for cultivation. "Slash and burn" is a farming method used in many countries, especially in Southeast Asia and Africa, where trees are cut and burnt to expand arable land while enriching its soil with ashes and nutrients, explain authors. "Controlled fires are also used deliberately on a much larger scale in commercial agriculture, clearing the ground for palm oil plantations in Indonesia and cattle ranches in Brazil — and these fires frequently run out of control. In the Brazilian Amazon, fires are also part of a pattern of increasing encroachment into public and Indigenous Peoples' lands," the report states.

What can be done?

According to the research team, fires are a critical global issue that needs urgent global solutions. To have any chance of restricting the rise in global temperatures to 1.5°C in line with the Paris Agreement, more needs to be done to cut carbon emissions from forest fires, by governments, businesses, communities, and individuals alike, says experts. The best place to start is to focus on the causes and the factors driving them, they emphasize.

"We need to raise climate change ambition worldwide, improve Paris Agreement accounting for emissions from 'non-anthropogenic' fires, reinvest in prevention, halt deforestation,clarify governance, coordinate policies, use a science-based approach to risk and interventions, and bring businesses on board. Wildfires are a global problem and need to be prioritized in public policy and health policy," the authors recommend.

The fire maps show the locations of actively burning fires **around the world** on a monthly basis, based on observations from the Moderate Resolution Imaging Spectroradiometer on NASA's Terra satellite. The colors are based on a count of the number (not size) of fires observed within a 1,000-square-kilometer area.

INDIAN SCENARIO

India is no different when we speak about forest fires. India has witnessed a rapid increase in the number of forest fires. Forest fires in India in the past has caused a large amount of destruction leading to loss of life and cultivable land. Here, in this article, we have provided some statistics that will show that the issue of forest fires demands attention and government must take some steps to ensure that fires can be contained before large-scale damage occurs.

Forest fires in India

Every year we have a brief spotlight on the forest fires in the news channels and newspapers. After that, there is not much ado about it anywhere. Recently, when there was a fire in Bandipur forest, it was making headlines for quite some time and of course the gruesome Theni forest fire, which claimed a lot of lives and reduced hectares of vegetation to ashes. After a couple of months, we forget these incidents.

2019 Bandipur forest fires

In February 2019, massive forest fires broke out in numerous places across the Bandipur National Park of the Karnataka state in India. The National Remote Sensing Centre of the Indian Space Research Organisation (ISRO) carried out an assessment of the total area affected by the fire. On 25 February 2019, it estimated the extent of burnt area to be about 10,920 acres in the last five days since 21 February 2019.

Though the major fire was brought under control by 24 February 2019, some areas continued to burn. The fire also spread to Mudumalai forest range in Tamil Nadu, causing damage in around 40 acres. The Tamil Nadu forest department took action to ensure that it did not spread further.

Background

On 21 February 2019, wildfire broke out in the Bandipur Tiger Reserve. Unlike in previous years, this is the first time the wildfire in Bandipur flared up earlier due to the sudden climatic change and rapid growth of dry grass and Lantana. Officials reported that there appeared to have been no deaths of the larger mammals in the Park, such as bison, elephants, leopard and tiger.

Over 10,000 acres of forest in Bandipur area was destroyed. With the fire spreading rapidly due to winds, the authorities closed the Gundlupet-Ooty National highway and safaris in Bandipur National Park were also cancelled. Strong winds were making the job of firefighters, forest staff and volunteers more difficult.

The fire that destroyed Kundakere Range, spread to Barakatte and Guddakere, then to the Himavathi Gopalaswamy Hills. It also destroyed forests in Jarkal Kere and Gowri Kalu hills. Many smaller mammals and reptiles were killed, along with thousands of trees. There had been no forest-fire incidents noted in Bandipur National Park in the previous two years.

Firefighting

Two Mi-17 Indian Air Force helicopters were deployed on 25 February 2019 afternoon to douse the fire, which had been raging for three days in the Bandipur Tiger Reserve. The operation on 25 February 2019 was primarily limited to the Gopalaswamy Hills Range in Karnataka's Chamarajanagar district. The other two ranges that were seriously affected were Kundakere and Bandipur. The helicopters were deployed after Karnataka state Chief Minister HD Kumaraswamy called on the Indian Air Force for help. One proceeded towards Karadikal hill ranges while the other fought fires in the general area of Chammana halla top.

Bolagudda and Kanive temple area were also subject to helicopter fire control A total of 10 sorties were flown, dropping approximately 30,000 litres of water.

Figure : Forest department crew and local people helping to douse the Bandipur forest fire on 24 February 2019.

Damage assessment

Satellite images shown the extent of the damage caused by the wildfire. The National Remote Sensing Centre of the Indian Space Research Organisation (ISRO), Hyderabad, released a report based on Sentinel-2 satellite data, which reported that 15,443.27 acres of forest area damaged by the fire between 23 February 2019 and 25 February 2019 in Bandipur forest region alone. Nature Conservation Foundation's Wildlife Scientist Mysore Doreswamy Madhusudan estimated that 17,000 acres would have been affected since the National Remote Sensing Centre excluded the damage in administratively part of Bandipur (Revenue land).

Number of casualties due to fire has not estimated yet. Wildlife officials say that small animals, reptiles, which are slow-moving, would have borne the brunt, while bigger mammals like leopards and tigers would have fled to the Bandipur Tiger Reserve and taken refuge in nearby areas.

Investigation

Karnataka's top forest official confirmed that an "act of sabotage" had caused the blaze. Forest officials have arrested two shepherds for allegedly setting fire to the forest. Fearing tiger attacks on their cattle in Bandipur, on 22 February 2019 they had sparked the fire to chase away a tiger which was spotted in the area for a month or so, leading to massive fire, which destroyed thousands of acres of Bandipur forest area. The accused were booked under Indian section 27, 29, 30, 31, 50 and 51 of the Wildlife Protection Act, 1972.

2016 Uttarakhand forest fires

The **2016 Uttarakhand forest fires** were a series of widespread, damaging wildfires that took place in Uttarakhand, India between April and May. The fires were caused by a heatwave that spread across Uttarakhand and were the worst recorded in the region with a reported 4,538 hectares (11,210 acres) of forest burnt down and seven people dead.

Officials detected nearly 1,600 total fires which were brought under control by 2 May, and as rain fell the following day, it reduced the impact of the wildfires. An initial report on 4 May noted that 3,500 hectares (8,600 acres) had been destroyed by the fire.

History and causes

Forest fires caused by humans in the Himalayan state of Uttarakhand have been a regular event. Since the formation of the state in 2000, more than 44,518 hectares (110,010 acres) of forest land has been lost to fire.

Some fires during 1921, 1930, and 1942 have been associated with popular movements against the then British government's forestry policies and as a call for independence. Most of these fires occur in the chir pine (*Pinus roxburghii*) zone which is prone to fire.

But the pine itself is resistant to the fire. Instead of harming them, these fires help in their regeneration by reducing coverage of broad-leaf trees, which leads to an increase in the land covered by chir pine and oak. The broad-leaved forest is dominated by oak species, including Banj oak (*Quercus leucotrichophora*). Resin is harvested from these pines, and fires are often intentionally set to remove the carpet of pine needles that come in the way of harvesting activity. In June 1981, a blanket ban was introduced against the felling of standing trees 1000 meters above sea level, a ban that prohibits the removal of pine trees in the areas where they are invading. Apart from accidental fires, several causes and motivations for setting fires have been noted which include encroachment on forest lands and concealing illegal timber extraction. Major illicit timber trade relies on pine forests since it is often used for construction.

Figure : Burnt Chir Pine

A major cause for the fires was attributed to the record high temperatures and severe lack of rainfall due to a dry winter. The Uttarakhand forest fires had also come as India suffered one of its worst droughts in years with the government noting over 330 million were affected by water shortages.

Early 2016 events and controversy

The forest fires began on February 2, 2016, and went initially unnoticed for several weeks, while Parliament was in a state of outrage over the presidential rule which had been imposed on Uttarakhand Consequently, these fires went largely unchecked and spread from the pine forests on the slopes of the sub-Himalayan region to the oak and broadleaf forests which resulted in heavy clouds of smoke that changed the usually blue skies to grey. This smoke affected tourism within the state and fearing bad publicity, the locals did not report the fires.

The conservationist Ela Smetacek started a public awareness campaign to highlight the damage from the fires calling on the Indian government to address the issue. Her Facebook post was shared over 58,000 times as it showed the entire hillsides ablaze with walls of flames engulfing even the tallest trees.

The campaign gathered widespread support from social media users and Indian media which led to a series of claims and counter-claims between the environment minister, the conservationist, and other activists ensued, as Smetacek blamed the timber and builder mafia for the purposely-lit fires, stating that 1,300 fires could be seen from NASA satellite pictures and that an arson attack across the state was evident. The post demanding the fires be recognized as a national emergency, and the need of immediate deployment of NDRF teams, led to the government taking action, deploying the National Disaster Response Force and making use of Indian Air Force Mi-17 helicopters fitted with "Bambi buckets" to douse the fires with water.

Seven fatalities were reported as of 4 May 2016. The Pinegrove School, a boarding school at Kasauli, was evacuated as fires had reached its compound walls. Tourism and wildlife at the Jim Corbett National Park and Rajaji Tiger reserve regions were affected. Various other locations in these Himalayan states of Uttarakhand and Himachal Pradesh are tourist attractions in summer, and they now face heavy air pollution. The forest fires also disrupted the functioning of the Kalka–Shimla Railway line. On 3 May, the forest department estimated the monetary losses at approximately 29 lakh (US$41,000). The estimate is based on the standard rule book followed by the department with empirical formulae for such calculations. Various ecologists and environmental activists have disregarded the estimate, noting that it does not take into account ecological and wildlife losses. They added that these fires have also destroyed vegetation which holds rainfall, which might result in floods in the monsoon season.

Scientists of the Govind Ballabh Pant Institute of Himalayan Environment and Development formed a team to survey the area and study the effects of the fires on the melting of glaciers. Ecologists suggested that clearing of forest floors of the fallen pine leaves, which are readily combustible, should be undertaken by the forest department as well as locals on a grassroots level to prevent such major fire outbreaks. Production of biomass briquettes from these pine needles should be promoted, serving as a source of fuel as well as a solution to prevent wildfires. On 30 April, the National Disaster Response Force was deployed for rescue operations in the Kumaon and Garhwal areas. Around 6,000 personnel from the Forest Department were deployed. As of 30 April 2016, 922 incidences of fire were reported affecting around 2,000 hectares (4,900 acres) of forest area. The Mi-17 helicopters of the Indian Air Force with "Bambi buckets" were used to douse the fires with water. However, the heavy smog that developed in the area affected the aerial operations of the army. The affected area later increased to around 3,500 hectares (8,600 acres), extending into Himachal Pradesh. The regions saw rainfalls on 3 May ranging from 7mm to 11mm, which helped in controlling the fire.

Soaring temperatures reignited the forest fires on 18 May 2016 covering an area of nearly 180 hectares of green land spreading over 111 districts. Rangers and Divisional Forest Officers rush to the spot and extinguish the fire as soon as possible to reduce the damage done. This was the result of the temperatures of Uttarakhand which were about 4 to 5 degrees higher than the average temperatures at this time of the year. According to government sources, the forest fires combined have destroyed nearly 4,048 hectares of land, in 1,857 incidents.

Subsequently, the World Bank granted 500 crores rupees (US$75M) to the Uttarakhand forest department to research the cause of the forest fires, which was later increased by 600 crores (US$90M).

The Indian central government later announced a series of funds amounting to over 50,000 crores (US$7.5B) to go towards afforestation across the state of Uttarakhand and the country.

Figure : Vegetation destroyed in a forest near Dhaulinag Temple in Bageshwar district

Controversy and long term impact

Reports were published in the local media which variously blamed both the local people and timber companies for the forest fires. Environmental journalist Bahar Dutt blamed the television media for their "Delhi centrist world view" that disregarded the historical and ecological natural causes of fire, due to the high density of pine trees that increases the fire hazards. Various tourism-related associations suggested that the exaggerated representation of the fires by the media could reduce tourism, a major part of the state economy. Part of a team of conservationists and biologists who studied the private and state forests since 1950, Smetacek maintained forest fires did not occur naturally in the geographical topography of the region concerned. Natural fires are not seen anywhere in Indian forests, she added. The introduction of pine trees by the British caused a more volatile fuel to the (man-made) fires.

Fires directly destroy biodiversity and have indirect long-term impacts, including the encouragement of fire and pioneer species. It has been suggested that the dark carbon dust emitted by the fires deposited on Himalayan glaciers could hasten the glaciers' melting. This could affect the hydrology of the rivers that are a source of water for human populations in northern India. The average temperature of northern India saw an increase of 0.2 °C.

CHAPTER 5

FOREST FIRE MANAGEMENT

Introduction

Thousands of hectares around the globe destroyed by forest fires every year causing tragic loss of houses, properties, lives, fauna and flora. Forest fires are a great menace to ecologically healthy grown forests and protection of the environment. This problem has been the research interest for years, and there are a number of solutions available to resolve this problem.

A forest fire is unenclosed and freely spreading combustion that consumes the natural fuels of a forest—i.e., duff, grass, weeds, brush, and trees. Forest fires occur in three principal forms, the distinctions depending essentially on their mode of spread and their position in relation to the ground surface. Surface fires burn surface litter, other loose debris of the forest floor, and small vegetation; a surface fire may, and often does, burn taller vegetation and tree crowns as it progresses. Crown fires advance through the tops of trees or shrubs more or less independently of the surface fire and are the fastest spreading of all forest fires. Ground fires consume the organic material beneath the surface litter of the forest floor; ground fires are the least spectacular and the slowest-moving, but they are often the most destructive of all forest fires and also the most difficult to control.

A forest fire does a number of specific things. First, and perhaps most obviously, it consumes woody material. Second, the heat it creates may kill vegetation and animal life. In most fires, much more is killed, injured, or changed through heat than is consumed by fire. Third, it produces residual mineral products that may cause chemical effects, mostly in relation to the soil. The lethal temperatures for the living tissues of a tree (i.e., the phloem and cambium, which are located under the bark) begin at 49 °C (120 °F) if exposure is prolonged for one hour.

At 64 °C (147 °F) death is almost instantaneous. The ignition temperature for woody material is approximately 343 °C (650 °F), with a flame temperature of 870–980 °C (1600–1800 °F).

Forest fires seldom occur in tropical rainforests or in the deciduous broad-leaved forests of the temperate zones. But all coniferous forests, and the evergreen broadleaf trees of hot, dry zones, frequently develop conditions ideally suited to the spread of fire through standing trees. For this, both the air and the fuel must be dry, and the fuel must form an open matrix through which air, smoke, and the gases arising from combustion can quickly pass. Hot, sunny days with low air humidity and steady or strong breezes favour rapid fire spread. In coniferous forests the resinous needles, both living and dead, and fallen branch wood make an ideal fuel bed. The leaves of evergreen broadleaf trees, such as hollies, madrone, evergreen oaks, and eucalyptus, are coated in inflammable wax and blaze fiercely even when green. Once started, fire may travel at speeds of up to 15 kilometres (10 miles) per hour downwind, spreading slowly outward in other directions, until the weather changes or the fuel runs out.

Well over 95 percent of all forest fires are caused by people, while lightning strikes are responsible for 1–2 percent. In some countries the setting of fires for clearing cropland is an integral technique of agriculture. In other areas forest fire prevention, including public education, hazard reduction, and law enforcement, consumes a considerable amount of time and money. The two basic steps in preventing forest fires are reducing risk and reducing hazard. Risk is the chance of a fire's starting as determined by the presence of activity of causal agents, most likely human beings. Hazard is reduced by compartmentalizing a forest with firebreaks (alleyways in which all vegetation is removed) and reducing the buildup of fuel (litter, branches, fallen trees, etc.) by controlled burning. In the United States the Forest Service devised a National Fire-Danger Rating System, which is the resultant of both constant and variable fire danger factors that affect the inception, spread, and difficulty of control of fires and the damage they cause.

Effective fire control begins with a field survey and map to identify the areas at risk, delineate them, and define and improve the barriers or firebreaks that may limit fire spread. Natural barriers include rivers, lakes, ridge tops, and tracts of bare land. Artificial barriers can be roads, railways, canals, and power-line tracks, but usually extra firebreaks must be cut to link these and provide wider gaps that fire cannot readily jump. Belts of land from 10 to 20 metres (33 to 66 feet) wide are cut clear of trees or left unplanted when a new forest is formed. Sometimes the soil is left bare and cultivated only at intervals to check invasion by weeds. Usually it is sown with an even crop of low perennial grasses or clovers and kept short by mowing or grazing. This checks soil erosion, provides an evergreen fireproof surface, and allows access on foot, by car, or in an emergency by fire-fighting trucks. Surfaced roads, serving also for lumber haulage and access for recreation, are of critical importance in fire fighting. Signposts are needed to guide fire crews unfamiliar with the woods and to mark water supplies and rendezvous points.

Detection is the first step in fire suppression. Many countries have organizations of trained professionals to detect and fight fires; others rely on volunteers or a combination of the two. Tower lookouts are the mainstay of nearly all detection systems, although the use of aircraft and satellites has modified this view in countries with an advanced fire control program. Fire surveillance is essential during seasons of high risk. Towers are set on hilltops where observers equipped with binoculars, maps, and a direction scale determine the compass direction of smoke and notify the fire control base via telephone or radio. If a fire can be seen from two or more towers, its precise position is quickly determined by mapping the intersection of cross bearings. Aircraft are used to detect fires and to carry out reconnaissance of known fires. Aerial surveillance has probably been most successful in detecting lightning-caused fires and is most often employed in areas of relatively low-value lands and inaccessible areas. An aircraft is essentially a moving fire tower, and the problems of detection that apply to a tower also apply to an aircraft; however, new developments in remote-control television, high-resolution photography, heat-sensing devices, film, and radar make fire detection by aircraft and satellite more efficient and location more accurate. Satellites provide a rapid means of collecting and communicating highly precise information in fire detection, location, and appraisal.

Once a fire has been detected, the next step is fire suppression. The first job is to stop or slow the rate of spread of the fire, and the second job is to put it out. The aim of suppression is to minimize damage at a reasonable cost. This does not necessarily mean the same thing as minimizing the area burned, but it is a major goal. Suppression is accomplished by breaking the "fire triangle" of fuel, temperature, and oxygen by robbing the fire of its fuel (by physically removing the combustible material or by making it less flammable through application of dirt, water, or chemicals); by reducing its temperature (through application of dirt, water, or chemicals and partial removal or separation of fuels); and by reducing the available oxygen (by smothering fuels with dirt, water, fog, or chemical substances).

The great majority of all forest fires are contained by professional fire fighters equipped with numerous hand tools (spades, beaters, axes, rakes, power saws, and backpack water pumps). Trained fire crews with light, hand equipment can be carried quickly to a fire by truck, delivered by helicopter, or even dropped by parachute. When necessary, large machines (bulldozers or plows) are used to clear openings, or firebreaks, which stop the spread of the fire. This requires clearing surface and sometimes aerial fuels from a strip of land and then digging down to mineral soil to stop a creeping or surface fire. A control line can also be established by directly extinguishing the fire along the edge or by making fuels nonflammable. In some cases a backfire may be deliberately set between the control line and the oncoming fire to burn out or reduce the fuel supply before the main fire, or head fire, reaches the control line.

Water is the most obvious, efficient, and universal fire extinguisher, but large-scale use of water in fire fighting is limited because it is usually in short supply and application methods are not adequate. For these reasons other materials have been tested for persistence and efficiency in putting out fires. Wetting agents change the physical characteristics of water to increase its penetrating and spreading abilities. Retardants, such as sodium calcium borate, reduce the flammability of wood and therefore its rate of burning. Foaming agents in powder or liquid form can greatly increase the mixture volume and thereby cool, moisten, and insulate the fuel.

Aircraft can quickly carry in water and other chemicals to be dropped or sprayed on the fire. A method developed on Canadian lakes is to fill the floats of a seaplane with water, which is done as it skims the lake on takeoff, and to discharge this through nozzles over the fire.

Historically, the prescribed use of fire in forestland management has been approached with understandable reluctance by many foresters and wildland managers. Yet, fire has a place in the management of particular ecosystems. The decision to use fire is usually based on a balancing of pros and cons—i.e., damage, possible or expected, must be weighed against benefits. Under proper circumstances, prescribed burning can be used to prepare seedbeds for natural germination of most tree species, to control insect and disease infestations, to reduce weed competition, to reduce fire hazard, and to manipulate forest cover type.

Forest fire management

Forest and savanna fires release large amounts of carbon into the atmosphere in a short amount of time. Ground fire is often a natural feature of healthy ecosystems, but uninformed past human interventions have left forests and savannas more susceptible to catastrophic fires. Better managing how and when forest and savanna burns can prevent excessive loss of carbon into the atmosphere.

This climate mitigation pathway involves three discrete types of fire management:

- Prescribed fires in fire-prone temperate forests to reduce emissions of biomass from the historic rate of wildfire losses.
- Prescribed early-season burns in the savanna to avoid higher emissions from late-season burns.
- Fire control practices in Amazonian forests to avoid unintended fires that degrade the forest (such as fire breaks at forest edges).

The Numbers

Worldwide, avoided fire impacts could be applied across an area of 1 million hectares per year, an area larger than the size of Puerto Rico.

Fire management would prevent the release of up to 127 million metric tons of carbon dioxide equivalent per year (MtCO2e/year). That's comparable to the emissions from 27 million passenger vehicles per year.

The Challenges

Improving fire management will require overcoming some challenges. Implementing prescribed fires and fire breaks will require an initial investment of upfront capital. And because fire management is not a one-time intervention, there will be ongoing annual costs.

Conservationists must also consider that prescribed fires will involve increased initial emissions in order to avoid future larger emissions. In addition, as it is not possible to know the precise location of future wildfires, fire management must occur routinely across a large area.

Moving Forward

As a result of these challenges, we estimate that fire management is a relatively higher-cost pathway. However, this pathway involves low-tech, well-understood fire management practices, which could be implemented immediately if the funds were made available.

Partnerships with local communities will be an important part of moving this pathway forward. Incentives and capacity building will be helpful in improving fire management, in storing more carbon and improving local livelihoods.

Case study

In countries such as Australia, intense, uncontrolled wildfires late in the dry season contribute significantly to annual greenhouse gas emissions. The West Arnhem Land Fire Abatement (WALFA) project aims to shrink that contribution through more frequent, but less intense, prescribed burns.

The project is a partnership between Australian Aboriginal Traditional Owners and Indigenous Ranger Groups, Darwin Liquefied Natural Gas, the Northern Territory Government and the Northern Land Council. WALFA pays Indigenous Fire Rangers to conduct traditional strategic fire management across 2.8 million hectares of Western Arnhem.

In its first three years, the project realized a 38 percent reduction in greenhouse gas emissions relative to the 10-year project baseline.

Notably, the project demonstrated accounting methods for quantifying emission reductions resulting from fire management practices in a fire-dependent savanna system. The project has also proven the viability of engaging indigenous groups in traditional fire management activities in order to reduce fire emissions.

Improved fire management practices will be particularly important in African savanna ecosystems and the frontier forests of the Brazilian Amazon.

Integrated forest fire management

Once highly resistant to fire, significant areas of tropical moist forests have gone up in flames in recent years. The impacts can be devastating: tropical forest fires kill people, destroy homes, livelihoods and wildlife, and pollute water catchments. Smoke hazes from tropical forest fires create health and navigation hazards on a regional scale. The huge fires that burned in Borneo in 1997 and 1998 rang alarm bells among tropical forest managers and fire experts. Heavy rains finally extinguished the fires.

Integrated forest fire management (IFFM) comprises a systematic approach to forest fire management. It encompasses both the traditional efforts of fire prevention and fire suppression as well as the use of prescribed fire as a tool, community involvement, and forest law enforcement.

IFFM underpins the ITTO Guidelines on Fire Management in Tropical Forests, which were published in 1997. These provide a step-by-step process by which tropical countries can analyze their fire management situation and develop workable programmes to deal with it. Written to be broadly applicable throughout the tropics, the guidelines contain seven categories of issue important in any fire programme; within each of these are principles known to affect fire management, and for each principle there are recommended actions. Those using the guidelines must evaluate the local situation and decide if the recommended action should be applied as it is, modified to fit, or rejected as inapplicable in the circumstances.

Forest Fire Management Strategies Around the World

The objective of fighting a forest fire is the prevention of damage to people, property and assets. In addition it significantly contributes to environmental protection. Fundamentally, the protection and safety of the operational fire fighting force is of prime importance. Necessary fire fighting measures, which could put the rescue personnel in danger, should be limited as much as possible.

If a forest fire is first reported to fire service headquarters, they notify the rescue services and inform the forest authorities as well as forest owners.

By means of the modern sensory equipment used in forest fire early warning systems, forest fires can be quickly and accurately located. This head start can be decisive in the resulting size of a forest fire. For about 75% of forest fires in Brandenburg, 20 minutes elapses between detecting the fire and starting to put it out.

Local fire brigades extinguish most forest fires at their first attempt. Weak points in the alarm system or a breakdown in cooperation between the fire service and the forest authorities can have an impact on the size of the blaze and the amount of damage caused. Therefore regular forest fire drills, in which all the actors get to know each other and practise working together, are an essential part of forest fire fighting and prevention.

Operational Headquarters

If a fire cannot be extinguished at the first attempt then additional personnel and resources must be notified and brought to the site of the fire. If the extent of the fire makes it necessary to split the fire fighting forces into divisions, then an operational headquarters is established. The operational manager is the fire chief who is the first to arrive at the scene.

Fire fighting is coordinated through the operational headquarters (emergency services centre or operations centre) from a central location (command vehicle or previously determined space), which is not directly located at the endangered site. The manager directs all actions to combat threats, in particular the effective deployment of fire fighters at often unknown locations. Hence, the manager must quickly ascertain and evaluate the situation. S/He has to be able to rely on information and recommendations from the operational headquarters. The relevant local forest official must also be part of the operational headquarters team, as s/he has precise knowledge of the forest area.

The operational headquarters has the following duties:

- Estimate, control and evaluate the state of the fire, fire development and fire fighting measures as well as present recommendations on how to proceed to the operational manager
- Position fire fighters and resources according to the operational manager's commands
- Install necessary means of communications
- Ensure the provision of operational forces, the timely relief of personnel, medical supplies
- Prepare site plans and document issued commands, notifications etc.
- Set up barricades
- Detain at risk people

Strategies to fight forest fires

In contrast to building fires, the priority in forest and wild fires is to prevent the fire from spreading. In the rarest of cases, sites can be completely extinguished. The fire fighting strategy predominantly aims to contain the fire.

Type of forest fire

Forest and wildfires typically form roughly in the shape of an ellipse. Burning only occurs in the area of the perimeter of the ellipse. The fire flanks found to the right and left of the wind direction spread noticeably slower than those in the direction of the wind – the so called fire front or hot spot. If the wind changes direction, the flanks can quickly become the fire front.

It is important to assess fires on terrains such as mountainous areas. Along with the danger that fire personnel could fall, fire fighting on slopes and in valleys demand particular expertise. Generally a forest fire in a mountainous terrain spreads uphill very rapidly. This is caused by the upward flowing hot air and warm thermals. Higher growing, stunted vegetation is readily desiccated by the approaching fire. Furthermore, fires can be rekindled again and again by rolling and burning forest material.

The most important objective is to prevent the fire jumping from the ground into the tree canopy and hence starting a crown fire. It is possible to fight flame lengths up to 1m above the ground with fire beaters. Fire perimeters with flames the height of a person (1-2m) are fought with water jets along with backpack sprayers and shovels. Putting out flames higher than 2-3m is not possible using hand held equipment. With flames greater than 3m there is a danger of an aerial fire and therefore an increased risk of fire islands caused by firebrands. Due to the rising hot air, sparks of pinecones, moss, charcoal or birch bark tar can be lifted up and carried to ignite new fires up to 400m ahead of the fire front. In this manner flying sparks allow even wide strips of deciduous forest to be jumped.

Reconnaissance

In order to fight a forest fire successfully, a thorough survey of the site is a must. The key focus areas are determined according to the following tactical priorities:

1. Protection of people
2. Protection of animals
3. Protection of structures (buildings, streets, utility lines)
4. Protection of endangered or fast burning vegetation

When investigating the type of fire (crown, ground or surface fire), geographic and topographic particulars (main direction of spread of blaze, terrain specifics, entry and escape points for water tenders, wind direction and any expected changes) also play an important role. Conclusions can also be drawn about the fire by observing the column of smoke (already when driving to the site). The colour and form of the column of smoke provides information about the fire's behaviour.

Based on all this information a decision is made during the operational planning about how the fire will be fought. A defensive or offensive method of fire fighting is possible.

Offensive forest fire fighting

Common fire fighting practice in Germany is a direct offensive attack on a fire front by means of fire fighting crews, fire engines and/ or air tankers. An approach against the wind at the fire front is the most effective, but because of the difficulty in predicting the speed at which the fire is spreading, is not without risks. This method can only be applied at low flame heights. The danger to people is high if the speed of the fire and flame lengths has been falsely estimated. Increasing wind speeds and hard to manage topographical conditions can lead to fire services being enclosed by fire. Furthermore fire crews are exposed to the heat of the fire and smoke.

Defensive forest fire fighting

If a direct attack is not possible, because the flames are too high to tackle or the surface is contaminated with ammunition, then these blazes are fought defensively. By creating fire breaks (fire lines) or using existing fire resistant barriers (streets, paths) the fire should be able to be halted. Fire breaks can even be laid in advance allowing straight fire lines to be created. Along with the positive benefit of working without heat or smoke stresses, this method also has disadvantages. They include the increased workload and the danger posed to the fire crews by working without visual contact with the fire perimeter. Furthermore fires are able to jump these fire breaks and surveillance of the land lying behind them is required. It is important while fire fighting (defensive as with offensive) to monitor the surrounds and immediately put out any fire islands caused by sparks or firebrands.

Establishing and Safeguarding Firebreaks

Establishing firebreaks is generally only sensible when intensive aerial fires are to be stopped. In most cases the immense personnel and material effort required (clearing tools, staff etc.) can be more effectively used in an offensive procedure. However, their establishment can always be seen as a Plan B in order to safeguard the operational success of an offensive approach. Coordination with the relevant forest officials and forest owners must occur before their construction. Creating firebreaks usually requires using large equipment. Forestry and agricultural machinery as well as bulldozers or military armoured recovery vehicles can be used to undertake the scarification (exposing the mineral soil layer). Firebreaks are always created in two pieces with a vegetation free scarified strip and an observation area lying behind it. The scarified strips should be so created that they are twice as wide as the expected flame length.

Ammunition contaminated lands

Special rules must be adhered to when extinguishing fires on military training areas (current or former) and on former battlefields (e.g. WWII). Fire accelerants and explosions increase a fire's intensity. Close cooperation between the fire service and local and expert advisers like the army and forest personnel is important. Fire fighting on ammunition contaminated takes place at a safe distance or at strategically important points. In most cases these fires are fought indirectly or from the air. One waits until the fire reaches "non-dangerous" areas and then it is extinguished there. Areas contaminated with ammunition which are ablaze can not be driven over with conventional fire engines. Hence, specially protected fire engines are used in such fire fighting. Active military exercise areas have their own fire service and/or some armoured vehicles which can be deployed.

Five Ways to Prevent Forest Fires

Forest fires are a big issue in many parts of the world, when the temperatures rise and there is no rain in sight. The beautiful province of British Columbia that PerfectMind calls home has been facing some harsh forest fires, particularly in Williams Lake and the surrounding area.

The fire has been so detrimental that the entire city was evacuated, with fears of the blaze threatening homes and, more importantly, lives. While the evacuation order has since been lifted, Chilliwack's school district no. 43 took in evacuees who had nowhere to go—an amazing show of support. Forest fires aren't issues isolated to any state, province, or country.

While forest fires are sometimes a natural occurrence, human error and negligence is often the cause of many of the forest fires that we face. According to the National Interagency Fire Center, people cause an average of 61,852 fires each year. Wildfire prevention starts with the agencies that govern restrictions, but it ultimately falls to the awareness and actions of the general public. Below are five ways to help prevent forest fires.

It Starts from the Top

Regulating bodies need to be vigilant and preemptive with forest fires, in order to help curb their frequency. Those responsible for enacting a fire ban should be cross-referencing weather patterns with historical forest fire data. Prevention is key here and the message of a fire ban needs to spread quickly. Most cities are already equipped for this, as newspapers, radio, and local TV stations and programs disperse the information quickly. With the prevalence of social media and smartphones, it can also be useful to create a digital campaign that keeps the public informed.

Citywide Preparedness

For communities that live close to forested areas susceptible to forest fires, citywide preparedness is necessary for mitigating the damages inflicted. There are many existing programs, such as Fire Adapted Communities, Community Wildfire Protection Plans and Firewise, that help communities take individual and collective steps to protect themselves. These plans work to protect people, property, and natural resources, by emphasizing forest fire readiness and fire-resistant community design. Creating and enforcing the right regulations will help to contain fires and minimize damage, by reducing a fire's ability to spread.

Awareness and Education Is Key

For city dwellers who may feel detached from the issues brought on by wildfires, it can be hard to understand the importance of forest fire prevention. With many urbanites heading to the wilderness during the warmer months, it's important to raise awareness around the implications of a forest fire and how to minimize fire risks. Creating permits that people have to obtain before they can work in forestland, start campfires, or burn grasses, brush or debris lawfully can help mitigate the lack of awareness. While the basics of fire prevention may be common knowledge, there are many things that common sense alone won't tell you.

Responsible Behavior

While awareness is important, many people will have to make sure that it directly reflects in their actions. Part of this may include simple reminders, such as clear signs at a bottleneck for visitors of high-risk areas. Enforcing and managing the fines and penalties for violating any bans or regulations may also help reduce the willingness of individuals to engage in risky behavior.

As wildfires are a common problem throughout the globe, interacting with other forestry and parks and recreation organizations can reveal what has worked well, in other communities, and what has not. Regional think-tanks and governing bodies can help create a strategy for at-risk areas where each local population and fire prevention and suppression group can be involved.

Parks and Recreation Integration

Integrating fire prevention awareness into existing parks and rec programming, facilities, and parks can help spread the message. Having informative pamphlets at recreation facilities, touching on wildfire prevention in nature and forest workshops, and placing engaging and informative signs at parks are some of the many ways parks and rec organizations can help. Many organizations also have the ability to communicate to their communities quickly and more efficiently through various channels, such as social media networks, mailing lists, and newsletters. Helping spread the information can make a difference in avoiding wildfires.

Strategies Firefighters Use to Fight Wildfires

Most people know how to stop, drop, and roll. Some even know how to employ a fire suppression blanket and how to properly operate a fire extinguisher. But for the vast majority of people, fire is something you brag about being able to make out of two sticks, or something you cook dinner on top of, or just something you run away from—certainly not something you want to have to fight to put out. This is all probably for the best—battling a wildfire is no straightforward task. The men and women who do run toward an unexpected blaze have been well trained in strategies that can suppress the flames of a wildfire. Below are 10 of the techniques firefighters use to put fires out quickly, completely, and safely.

1. **CONTROL LINE**

   One of the most important components of wildfire suppression, *control lines* are simply the boundaries—natural or manmade—that firefighters employ to control how and where a fire spreads.

A rocky ridge or river can serve as a natural control line, or firefighters can establish a manmade one by, for example, clearing out an extended line of brush. Within this overarching definition, a *fire line* is when the barrier is scraped down "to mineral soil," a *scratch line* is a preliminary line built in a hurry, and a *wet line* is when the area has had flame retardant or water applied. And when you hear on the news that a fire is "X percent contained," this is generally what they're talking about—that some percentage of the fire's perimeter has a control/ fire line. Sadly, because fires can sometimes jump the barrier, this means even a 100 percent contained fire can start up again.

2. **BURNING OUT**

When establishing control lines, digging a small ditch and pulling up some plants isn't always enough. To create a sturdy, fuel-free barrier, firefighters may use small torches to burn the brush just inside a control line. A *burnout* is one of several ways to bolster a control line and further prevent a blaze from escaping the established boundaries.

3. **BACKBURN**

A *backburn* is similar to a burnout, but requires a slightly more sophisticated technique. Once a control line is established, firefighters may set a controlled blaze downwind of the main fire, just on the inside of the control line. Firefighters then push the new blaze back toward the main fire, burning up all the fuel that lies between the fire and the control line.

4. **FLANKING**

For a wildfire small enough to be snuffed out using a direct attack, firefighters may begin their assault on the blaze from behind. Starting from already burned earth, the firefighters will typically work their way around the edge of the fire to spray the flames as they make their way around the entire perimeter of the blaze.

5. **HOT SPOTTING**

Hot spotting is the term used to describe the extra attention given to the most active and dangerous portions of a wildfire. The crews fighting the fire size up the parts of the blaze most likely to spread and try to devise the best strategy for keeping these areas in check. Hot spotting may also involve diverting extra manpower to the task of stamping out embers and spot fires that blow or erupt from the hottest part of the fire.

6. **KNOCK DOWN**

Whereas hot spotting refers to the assessment of a fire's condition, *knocking down* is all about action. The knock down strategy is employed when firefighters decide that a certain hotspot needs to be suppressed immediately. To diminish the section of a fire deemed to have grown too hot, too active, or too large, fighters directly apply some combination of dirt, water, or retardant to that section.

7. **COLD TRAILING**

While a fire is being attacked from the front or side, other firefighters may be involved in *cold trailing*, the task of combing through already scorched ground in the wake of a moving wildfire. The point is to make sure no hot or glowing embers remain, since leftover coals can be blown around and flame up again.

8. **AERIAL ATTACK**

If significant exposed water sources are nearby, planes and helicopters can scoop up buckets of water and carry them to be dropped atop the blaze. The water is often mixed with a foam retardant before being dropped. The foamed water acts as a more effective barrier to the spread of fire and also insulates fuel that has not yet burned.

9. **FIRELINE EXPLOSIVES**

When setting control lines or firelines, firefighters may even use explosives to break up dense brush and fallen trees. Explosives can also be used to fell trees whose spread might help a fire jump across a control line. During a large or fast-moving fire, explosives are employed mainly for efficiency purposes, as they can save precious time when firefighters need to contain a fire quickly.

10. **MOP-UP**

It's called *mop-up* when firefighters go back and clean up along a completed control line. Mop-up consists of dousing any embers and spot fires that have made their way across control lines. It also involves protecting still-vulnerable fuels using a burnout (if they're permanently situated) or by simply moving them.

Global wildland fire and land management

Global wildland fire and land management encompass a wide

spectrum of situations and consequences. In many ecosystems, fire is a positive factor vital for plant health and survival while in others it can be a primary source of negative impacts such as biodiversity loss, ecological impacts, and social and economic consequences. Increasing wildland fuel loads, expanding wildland-urban interface areas, increasing human caused fire numbers, climate altered fire regimes prone to longer fire seasons, higher fire intensities, and greater fire severity of fire impacts are heightening overall fire complexity. At the same time, demands on and expectations of fire management response resources are rapidly mounting, risks to firefighters and fire management costs are reaching unprecedented levels, and aging equipment and infrastructure are becoming more difficult to upgrade or replace.

The future of wildland fire management cannot be predicted with a high degree of reliability but there is little doubt that we have entered a very transformative time. A review of and an adherence to the wildland fire framework of policies, strategic plans, land and resource management plans, and other guiding documents are important. This information framework varies across political boundaries and organizational missions but universally provides in-depth information and guidance to the wildland fire management program. It frames program planning and implementation and carries far greater value than ever before. It allows for greater flexibility, keeps pace with a dynamic situation, and embodies the state of the knowledge, the state of the art, and latest science and technology.

Fire policy has been quite responsive to changing situational dynamics. It has progressed to a point where it is the most comprehensive policy yet for wildland fire and decision-makers have more flexibility than at any previous time. Accepted strategies are more sophisticated and thorough and tactical spectrums span a wider range and implement multiple objectives; this is extremely important since a viable *one-size-fits-all* option does not exist.

Large-scale strategic planning is becoming more important and its value in providing direction and setting a national or even international vision for wildland fire management cannot be overstated.

National level strategic planning has been pursued in the United States over recent years and has produced the 2014 National Cohesive Wildland Fire Management Strategy, (http://www.forestsandrangelands.gov/strategy/). Many other geographic areas worldwide have also identified a number of strategic recommendations very similar in content to the USA Cohesive Strategy and have been presented at the 2015

International Wildland Fire Conference (http://www.fire.uni-freiburg.de/korea-2015.html).

Strategic thinking and planning must recognize and accept fire as a natural process necessary for the maintenance of many ecosystems and endeavor to reduce conflicts between fire-prone landscapes and people. The USA Cohesive Strategy states its vision for the next century as:

To safely and effectively extinguish fire, when needed; use fire where allowable; manage our natural resources; and as a Nation, live with wildland fire.

This vision is not only highly relevant, logical, and supportable in the United States, but is applicable in many countries around the world. It can in reality, be viewed as and has been called an international wildland fire management vision. Recommended necessary goals to achieve this vision have been provided in the documents referenced above, and can be grouped into six broad categories: fire management; fire prevention/suppression; science; technology/information systems; risk management; and cooperation and collaborative planning.

These goal areas include but are not limited to the following principles and core values:

Fire Management

- Restore and maintain landscapes:
- Wildland fire, as an essential ecological process and natural change agent, must be incorporated into the planning process and wildfire response.
- Actively manage landscapes across all jurisdictions to make them resilient to fire related disturbances in accordance with management objectives.
- Improve and sustain both community and individual responsibilities to prepare for, respond to, and recover from wildfire through capacity-building activities.
- Human populations and infrastructure are able to withstand a wildfire without loss of life and property.

Fire Prevention and Suppression

- Rigorous wildfire prevention programs are supported across all jurisdictions, cannot be reduced in favor of suppression activities, and constitute long-term efforts.

- Safe aggressive initial attack is often the best suppression strategy to keep unwanted wildfires small and costs down.
- Fire management programs and activities are economically viable and commensurate with values to be protected, land and resource management objectives, and social and environmental quality consideration. The principle of total fire extinguishment should be questioned in specific cases.

Science and Technology/Information Systems

- Fire management plans and decisions are based on the best available science, knowledge, and experience, and used to evaluate risk versus gain.
- Research supports increased scientific knowledge of biological, physical, and sociological factors.
- Scientific results and new management tools must be made available to managers in a timely manner and must be used in the development of land management plans, Fire Management Plans, and implementation plans.
- Updated data sets and advances in technology are necessary to improve decision-making processes.

Risk Management

- Sound risk management is the foundation for all management activities.
- Reducing risk to firefighters and the public is the first priority in every fire management activity.
- All jurisdictions participate in making and implementing safe, effective, efficient risk-based wildfire management decisions.

Cooperation and Collaborative Planning

- Local, state, tribal, national, and international organizations support one another with wildfire response, including engagement in collaborative planning and the decision-making processes that take into account all lands and recognize the interdependence and statutory responsibilities among jurisdictions.

The following figure shows how the principles and core values of the six listed goal areas can contribute to — and are essential to — an efficient and responsive wildland fire management program.

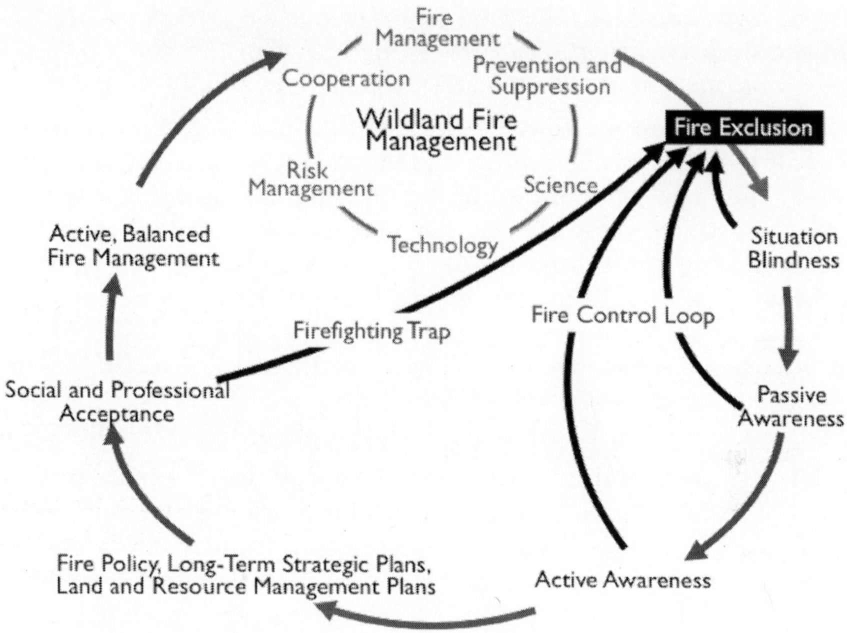

Figure 1. Wildland fire management efficiency dynamics.

Efficiency is maximized at the top of the chart in the small circle when all the goal areas are incorporated into fire management business. But when these activities receive diminished attention, efficiency decays. A *Situation Blindness* phase first occurs where it is not know there is a problem. Wildland fire management can move from a multi-faceted, dynamic program addressing multiple objectives with actions based on the best available science to a single focus program of fire exclusion. This occurred in the USA early in the 20th century and is attributed as one of several principal causes of altered fuel conditions and increased megafires.

Continuing on this pathway, the Passive Awareness phase occurs where it is known that there is a problem but it is either not known what to do or no actions are taken. Gifford Pinchot, the first US Forester, in 1899 in *"The relation of forests and forest fires,"* inferred the lack of understanding and scientific information in regard to the role of fire as an essential ecological process when he said,

"It is unfortunate that our acquaintance with the creative action of forest fires should be so meager, for only through a knowledge of this relation and through the insight which such knowledge brings

can there be gained a clear and full conception of how and why fires do harm, and how best they may be prevented or extinguished," and *"Remarkably little attention has been accorded to the study of forest fires as modifiers of the composition and mode of the life of forest."*

These statements reflect the lack of full understanding and that such understanding could improve management, but provides no action to remedy the situation. Continuing to move on from Situation Blindness progresses into the *Active Awareness* phase. Here it is clear that there is a problem and something needs to be done about it. During these first three phases, actions can easily turn back to a fire exclusion strategy through what can be called the *Fire Control Loop.* This is where actions are directed toward short-term fixes and represent passive approaches that place an over-reliance on past practices, processes, applications, experience, and state of knowledge.

To move forward, it is necessary to develop flexible policy, long-term strategic plans, and land management plans with clear objectives. Once this is done, a guiding framework exists and the program can progress into a stage of *Social and Professional Acceptance.* However, in the face of increasing fire complexity and negative impacts, even with a solid framework in place, it is easy to shift to an attitude that every wildland fire is an emergency and extinguishment is necessary. Resistance to change can be very strong at times and external pressures can come to bear and force a return to a fire exclusion strategy – the *Firefighting Trap.* In the USA in 1910, one of the most momentous fire events and loss of life occurred. Despite what Pinchot had offered about learning more about the natural role of fire, the significance of the 1910 fires fueled statements like: *"That the fire menace is a real one needs no emphasis,"* and *"The record shows a periodic recurrence of bad fires"* (F.A. Silcox. 1910. *Fire Prevention and Control on the National Forests*). This reaction how falling into the Firefighting Trap drove the program into a fire exclusion strategy under the "10 AM Policy" where **all** fires were to be suppressed as quickly as possible.

If thorough planning and preparation are completed and *Social and Professional Acceptance* occurs, an active, balanced fire management program can be achieved and implemented. Balance implies program focus on preventing wildfires; extinguishing fires where needed; using prescribed and natural fires where allowable and desirable; managing natural resources and restoring landscapes; conducting research and linking science and technology with management; incorporating risk management as an integral part of wildland fire management; gaining cooperation and participation by all involved organizations; and living with fire.

A well-planned course of action can make substantial and well-needed differences in the fire situation. Long-term strategic planning, patience, and commitment is necessary in order to affect needed changes. Strategies for wildland fire management must be proactive into the future and not just reactive.

Forest Fire Management in India

We cannot stress less on the inevitably of forests in the lives of human beings. It is imperative that we find strategies for maintaining the forest in a sustainable way, with all the benefits it can offer. Few countries have all the answers to all the problems faced, so international cooperation is needed. Loss of forest resources transcends national borders and affects the planet as a whole. In light of this, the roles of different agencies become vital to minimize any potential downside and maximize the upside.

Especially the global concern of wildfires across the globe. Wildfire will continue to affect the quality of the source water resulting in increased costs for care, repair and service. Forest and watershed managers and water sources, therefore, need to be well informed about wildfire impacts in order to establish mitigation plans to create water resource resistance to wildfire. They will collect detailed information about the extent and timing of post-fire impacts, proposed approaches to be thoroughly formulated and implemented. The bottom line, however, is that wildfire impacts should be integrated into regular forest watershed and water source preparation, safety, and operations.

With at least one in four people dependent on forests for their livelihood, India is losing at least Rs. 1,100 crore due to forest fires every year, says a new World Bank report.

Since independence in 1947, India has suffered a rapid depletion of forest resources. Some 75 million ha are officially classified as forest land, but according to the 1987 Forest Survey of India (Ministry of Environment and Forests, 1987), the actual forest cover is 64.2 million ha. This is equivalent to only 19.5 percent of the total land area, whereas the National Forest Policy (Ministry of Environment and Forests, 1988) sets a goal of at least one-third (approximately 110 million ha) of total land area under forest cover for ecological stability. Moreover, the existing forest resources are subject to severe biotic pressure, owing to the fact that with less than 2 percent of the forest area in the world, the country supports over 15 percent of the world's population and nearly 14 percent of its cattle.

Forest fires are a significant and increasing contributory factor in this degradation process, although the extent of total damage is widely disputed.

Causal factors

Most of the forest fires in India are deliberately set by small-scale farmers or landless rural people. In the northeastern parts of India, the practice of slash-and-burn shifting cultivation is the leading cause of forest destruction. The most heavily affected areas are Arunachal Pradesh, Assam, Manipur, Meghalaya, Mizoram, Nagaland and Tripura. Nationally, an estimated 4.35 million ha are affected by fire as part of shifting cultivation. This is especially significant in that with the growth of population and consequent decrease in the land/person ratio, not only are more areas subjected to shifting cultivation, but also the fallow period for regeneration has been reduced from the initial 30 years to two years.

Another of the most important causes of forest fires is related to the need for fodder for grazing animals. Although some 12.5 million ha of land is officially classified as permanent pasture or grazing land, most of this area is virtually devoid of grass. Therefore, most of the grazing requirement is met from forest area by setting fires to produce new flushes of grass in the dry season.

In central India, the production of tendu leaves (*Diospyros melanoxylon*) is another major cause of forest fire. Tendu leaves are rolled and filled with tobacco to make a cigarette known as *beedi*. Their collection and sale is an important secondary source of income. Collectors of tendu leaves set fires in the summer months to promote a better flush of leaves. The problem is compounded by the fact that tendu areas are leased for the collection of leaves on an annual basis; the purchasers are therefore reluctant to make long-term investments in crop improvement. The most economical and quick alternative is to set tendu areas on fire; these fires all too often then extend to other forest areas owing to negligence and carelessness.

Another non-wood forest product that contributes to forest fire is the mahua flower (*Madhuca indica*), collected by local people in north-central India to produce a popular beverage, or boiled with sal seeds (*Shorea robusta*) as a seasonal grain substitute. Mahua pickers burn the dry leaves under the trees to get a clean patch of floor to facilitate flower collection. While the intention is only to clear a small patch beneath a single tree, these fires often spread out of control. Since the collection of mahua flowers is done during the summer months, the hot dry weather aggravates the situation.

Effects of fire

Uncontrolled fires affect forest resources in a variety of ways. Regeneration is killed or dies back, thereby delaying the establishment of a new crop and extending the rotation. When newly planted teak plantations are bums, it is standard practice in Maharashtra to cut the young trees down to ground level; this stimulates a new vigorous shoot from the base, but at least one year's growth is lost. Fires are reported to damage seriously the regeneration of important tree species in sal forest (Maithani, Bhauguna and Lal, 1986). Chir pine regeneration is similarly killed or set back. Young eucalyptus plantations frequently require replanting and coppice regeneration dies back (or must be cut back) after fire.

Mortality may result from intense fires in older crops, although the trees develop thick bark that protects them. Eucalyptus appears to suffer more than the indigenous species and the effects of fires are apparent in reduced stocking per hectare and lower yields at maturity. In the Haldwani area, the actual yields obtained from pulpwood plantations, compared with theoretical yields derived from yield tables, indicate an average stocking of about 60 percent. However, it is not possible to say how much of this reduction is due to fire and how much is attributable to other causes.

Loss of increment results from the defoliation that accompanies fires and the consequent check to tree growth. Repeated burning leads to site deterioration, changes in soil nutrient status and accelerated erosion due to the destruction of the ground flora; these also reduce the rate of growth. No research appears to have been done in India on the growth losses attributable to wildfires. The only available evidence is a paper by Rawat (1949) on the results of controlled burning in the sal plantations of Bengal, which shows that fires significantly reduced the diameter growth of trees. Australian studies indicate that the volume increment of various species of eucalyptus is reduced after fires and that the effect persists for several years. The cumulative loss of annual increment depends on the severity of the fire, but generally lies in the range of one to three years' growth.

Timber quality is affected by scorching from the base of the tree, which damages the cambium, leading to defective butt logs. Fungal infection may occur through the damaged tissues and cause rot. The records of timber sold through the Ballarshah depot of the Forest Development Corporation of Maharashtra reveal that the average price received in 1984/85 and 1985/86 for fire-damaged logs was 9.8 percent lower than the average price of all logs sold. In the Nainital chir pine forests in Uttar Pradesh, resin tapping affects the yield of merchantable timber by damaging the lower part of the tree; the scars enable fires to bum into the heartwood and, in some cases, kill the tree.

These tangible losses may produce a knock-on effect in the form of a reduction in output and employment in industries dependent on wood as raw material. For example Chatterjee (1978) estimated that for sawnwood, each rupee's worth of output created Rs2.169 of employment and Rs11.994 of non-labour value added. Despite the fact that these multipliers were based on 1965 prices, they suffice to evidence the effect of fires on the wider economy.

No research appears to have been done so far on the potentially far graver intangible effects of forest fires in India. Not only do uncontrolled fires bum down the vegetation, but also the organic matter is adversely lowered, increasing the frequency of flooding and causing soil erosion. In addition, wildlife patterns and habitat may be disrupted. The situation is exacerbated by a lack of fire protection planning knowledge and incentive.

Modern forest fire control project

The scale of the damage and the waste of natural resources due to uncontrolled fires has thus brought the issue of forest fires to the forefront of forest policy. An important start was made in 1984 when the FAO/UNDP-assisted project in modem forest fire control was launched by the Government of India. The underlying thesis of the project is that fire damages and losses can be reduced considerably by using the basic components of prevention, detection, suppression and communication.

The project is active in two-areas which together are considered representative of the overall Indian situation - Chandrapur and Haldwani. Chandrapur (Maharashtra State) covers an area of 162600 ha with an undulating terrain consisting of natural teak forests and extensive teak plantations. It also includes the Tadoba National Park and is well served by communication and road links. The Haldwani project area (372700 ha) in Uttar Pradesh State is more diverse and includes mountains with forests of chir pine and sal natural forests. The flat Terai region has been converted largely to eucalyptus plantations for commercial pulp production.

Project activities have been concentrated in four key areas as part of an integrated fire management strategy:

Prevention

Physical prevention measures have concentrated on the construction and maintenance of fuel breaks in the mountainous terrain, and the use of mechanized equipment (tractors with plough) to construct fire breaks in the plains to protect eucalyptus plantations. These measures have proved effective from the standpoint of reducing fire losses and size of fire; they also provide safe access and areas for fire crews to make effective initial attacks on approaching fire fronts. Other preventive measures include prescribed burning of roadside fuel loadings, particularly elephant grass, especially in highly fire-prone locations.

Educational measures to increase awareness of the danger and damage due to forest fire have also been demonstrated by the project. Posters bearing a fire prevention message or fire risk meters have been installed through the divisions in the project areas, to increase public awareness of seasonal fire severity. These measures have been buttressed by a fire awareness week held annually. A survey of villagers' basic attitudes toward forest fires was conducted as a starting-point for developing closer cooperation with rural communities in fire prevention, as well as in detection and suppression. The results of the survey were encouraging in that they revealed that more than half the villagers questioned were aware of the harmful effects of forest fires.

Detection

A key element of a successful fire protection programme lies in the early detection of forest fires. In the Haldwani project area, a network of 32 fire detection towers have been strategically positioned and are staffed during the fire seasons with trained personnel equipped with binoculars and a radio for rapid reporting of observed smoke or fire. Statistics gathered in the two project areas during the fire season of 1987/88 indicate that although the frequency of fires was high throughout this period, the fires were controlled at a smaller size. Seasonal employment of villagers during the fire season ensures that an adequate level of labour operating under the aegis of trained ground personnel is readily available.

Suppression

The traditional method of combating forest fires - using fire traces and beating with branches - is clearly inadequate in dealing with present-day conditions. The project has effectively demonstrated the successful use of labour-intensive, manually operated hand-tools in reducing fire damage. In this sense, the modernization of techniques and methods is a matter not of computerized high technology, or imported equipment, but rather of a successful coordination of easily available resources and an understanding of values at risk.

The average size of fire in Chandrapur came down from 190 to 18 ha during the 1988/89 fire season. This was made possible because the fires were contained by a fast initial attack, aided by equipment provided under the project.

CLEARING A FIREBREAK to protect a teak plantation near Chandrapur

A considerable achievement of the project has been the initiation of local production of many items of fire-fighting equipment, including basic tools such as the Pulaski-McLeod rake, shovels, fire finders and protective clothing. The local production represents a significant saving of foreign exchange for India. For example, the fire finders had previously been imported at the cost of about Rs40000 (US$2350) each. Their manufacture in India costs only Rs3500-4000 (US$206-235). Although initial time estimates for tool production were inadequate, these tools are now being produced in sufficient quantities to meet the needs of the two states involved in the project, and increased production for national use is possible. Moreover, these items meet accepted international standards and production for export is possible.

INSPECTION OF A TRAINED FIRE CREW with protective clothing and locally produced tools

Mobile water tanks, manufactured in India for mounting on the back of lorries, have proved to be a very effective tool where there is good road access and water supply is adequate. Water scarcity is being eliminated by digging wells, and a water-holding area is being constructed in the Himalayas to supply water for mobile tankers in the project areas.

The catchment basin in the Haldwani area, designed so that a helicopter could scoop up water with minimum delay, was completed and ready for the 1988 fire season.

Local manufacture of tools has been combined with extensive training of forestry personnel, both through field demonstration and by training in the United States.

The project has also acquired and is testing an aerial system comprising a helicopter with a bucket, a fixed-wing aircraft, a newly created helipoint and trained staff to manage and pilot the aircraft. The system was in operation for the first time during the 1988 forest fire season.

Various demonstrations utilizing both aircraft have provided the project personnel with a working knowledge of aircraft capabilities, limitations and options available. A major issue for the project is to determine the cost/benefit ratio of aerial forest fire-fighting techniques in a developing country like India where the consideration of economies of scale is very important.

READY TO ROLL a mobile tanker

It appears that the costs of helicopter fire-fighting cannot be justified in areas such as the Chandrapur forest which have good road access. The system would, however, be useful in the mountainous terrain of Haldwani. The fixed-wing aircraft has been demonstrated to be useful in multiple roles in detection and transport of equipment for overhead fire-fighting, directing ground reconnaissance and law enforcement; however, absolute cost is a constraint.

DISCUSSING THE FIRE DANGER RATING SYSTEM developed with FAO assistance

The project has, moreover, demonstrated the financial viability of extending the modern fire control methods into other areas, by formulating a sequential series of steps, representing different levels of investment which can be implemented one step at a time. Step consists of providing basic fire detection and communication equipment and supplying the crew with proper hand-tools. Step 2 involves improved detection equipment and communication facilities, jeeps and water-handling equipment. Steps 3 to 5 progressively increase the investment in equipment and crew capability to control fire. By step 5 every state forest plan would incorporate fire management.

Planning tools

A standard fire-fighting methodology has been developed for a fire danger rating system suitable for the main fuel types. The system, which has now been approved by the Government of India for use on a national scale, provides data for routine daily operations and has also become a useful planning tool for scheduling prescribed burning, as well as serving as a data base for planning fire-fighting strategy during the fire season.

Further enhancement of the system could include an improvement in data retrieval, which at present is effected through the laborious, time-consuming process of manually researching data records.

A key element in planning a well designed, wide-based protection programme related to specific conditions is fire-reporting. A report form has been specially designed for this purpose under the project and is now operational in the project areas.

Properly applied and analysed, this report will strengthen the data available on forest fires-direct causes and physical conditions (wind speed, fuel type, etc.), duration or extent, detection sources, costs and types of suppression method and damage sustained, and so on. The review of cost data would be especially significant in that no standardization has been attempted so far in evaluating the damage caused by forest fires.

The emphasis laid on fire-reporting after the introduction of the project is generating more reliable data on the incidence of fires and area burnt. However, in the reports that were investigated, economic damage or suppression casts were frequently miscalculated, with considerable underreporting. A possible reason is that the reporting form may be too exhaustive or may overestimate the forest guard's ability to complete it accurately. An improvement would be to carry out this recording procedure in two stages: an initial fire report, providing the basic data on each individual fire incident, and a detailed fire report, including standardized calculations to assess and quantify tangible damages, to be completed later by a forest officer. Moreover, representative one-hectare models for the main forest composition could be prepared for the calculation of values.

An important planning tool has thus been the development of standardization for evaluating the damage caused by forest fires. The productivity changes due to the reduction of annual area burnt were estimated by using a computer modelling system for the Chandrapur project area. The model was used to generate projection of the future annual yields for the "with fire" situation and the change to those yields caused by the rotation length reductions of ten years for teak plantations and 20 years for the natural forest.

Evaluation of performance and future development

Over the past five years the project has demonstrated the effectiveness of modern fire control methods under Indian conditions. The forest fire control personnel have shown that the methods used will reduce the area burnt to a tolerable level of resource management. The effectiveness of the project is seen in that even though the number of fire incidents (reported) remained constant, the area burnt was reduced in Chandrapur from 15 percent in the base year of 1985 to 1.86 percent in 1989, and from 13 percent to 1.3 percent in Haldwani. An economic assessment carried out by FAO (Saigal, 1989) indicates that this reduction of burnt area, through its effect on rotation length and crop productivity alone, has increased the gross domestic product generated by the forestry sector by an estimated average of Rs9087 (US$535) per hectare in the Chandrapur teak forest and Rs1295 ($76) per hectare in the Haldwani sal forest. In addition the project has shown the viability of labour-intensive, indigenous hand-tools and the utilization of fire lines and control burning.

The demonstrated success of the techniques developed under the project has led to an extension of this project into the second phase to coincide with India's Eighth Five-Year Plan. In phase two, the project is being extended to ten states covering some 40000 km^2 of valuable productive forest equivalent to 75 percent of India's commercial forest.

However, economic assessment has highlighted certain areas that still need to be further strengthened. There is an urgent need to consolidate the data base and increase its reliability. Work studies need to be initiated to set the course for future fire planning. This should lead to the development of fire plans based on an assessment of when and where fires occur and the characteristics of the area; these factors would be necessary for an efficient extension of the modern fire-fighting methods into other parts of India. Moreover, the project has not yet had the time to generate data for making cost-effective comparisons between alternative combinations of personnel and equipment. However, for the present, investment should continue to focus on basic equipment and organization. In addition, a research programme to investigate the effect of fire on crop productivity on an all-India basis should be implemented as soon as possible by the Forest Research Institute of India.

Indian regulatory and prevention authorities

Fire in the forestland has been a part of India for aeons and has played an irreplaceable and an indispensable part in the ecosystem. From replenishing nutrients in the soil, regenerating new species and helping them grow to aid in maintaining a healthy forest life for all the kinds of flora and fauna. But fires only have positive effects till the time are occasional and not spun out of balance. And these out of proportion forest fires are not exclusive to India. It has indeed become a global concern. Just after independence, the government endeavoured to look into the forest disaster management and work for the conservation of the same. The first move in this direction was the creation of the National Forest Policy in 1952, when the former Ministry of Food and Agriculture, Government of India, declared a Forest Policy for State Forest Management in India. The Indian Constitution in Directive Principle of State Policy [Article 48(A] and fundamental duty [51(A)(g)] provides for the conservation of forests and wildlife under the Forest Policy.

Fighting forest fires is even more essential in India and it has set ambitious policy objectives for the enhancement of its forest sustainability. As part of National Mission for Green India under India's National Action Plan on Climate Change, the government has committed to increasing forest and tree cover by 5 million hectares and to improving forest quality on another 5 million hectares. To fulfil the National Forest Policies aim of achieving 110 hectares of forest cover, the Government had to strengthen the country's forest fire management policies too. It started formulating comprehensive and result-oriented programmes and initiatives to overcome the age-old issue of forest fires. Thus it started taking assistance from international environmental organisations like FAO for the Technical Cooperation Programme(TCP). This study began in May 1995 till December 1995. Furthermore, in 2003 the Indian version of US's Incident Command System(ICS) was adopted which was looked into by the National Disaster Management Authority(NDMA).

Keeping in view the severity of forest fire, the existing organizational structure, the Forest Protection Division, headed by Inspector General (IG) level officer looks after the forest fire management work at National level with the cooperation of National Disaster Management Authority (NDMA), Disaster Management (NDM) Division of Government of India, National Institute of Disaster Management (NIDM), International organizations like FAO, Forest Survey of India (FSI), Forest Research Institute (FRI) and other regional offices of MoEF in the country. The Ministry of Environment and Forests(MoEF) has six regional offices in the country acting as coordinating offices with the state forest departments. The forest fire control activities are looked at the State level after standard forestry staff, who are still overburdened. It'll be advisable if a separate department or a subsidiary of the department is established at the headquarters of the state forest directed by the Chief Conservator of Forest level officer.

Policies and laws to prevent forest fires

Forest fires have created havoc in different countries varying from Australia to California to Siberia and the list goes on. While it disrupted the normal working of humans, claimed their lives and gave them serious troubles, this is considered as a payback to them for intervening in the ecological balance of forests. Humans are considered to be responsible for maximum forest fires in the world. Therefore, it becomes quite crucial to keep a check on human activity. For the same reason, the Ministry of Environment and Forests had a draft to make provisions for managing the forest fires in the country with the National forest policy 2018. It's principle act being the one passed in the year 1988.

India's Ministry of the Environment has revealed a draft of the new National Forest Policy proposing to limit policies and programs that conflict with forests covering sharp slopes, river catchments, lakes and reservoirs, geologically vulnerable terrain and other ecologically critical areas. The draft legislation includes steps to protect habitats from forest fires, including:

1. Charting and surveying vulnerable areas;
2. Strengthening and developing early detection systems;
3. Remote Fire Detection technology; and
4. Enhanced Public engagement.

Numerous initiatives have been adopted to enhance forest fire management. Data is to be extracted via satellite and Fire Danger Rating System has been adopted and included in the policy to develop, strengthen and increase the efficacy of the Early Warning Fire Detection System. Forest Survey of India (FSI) works toward this Rating System in collaboration with the National Center for Medium-Range Weather Forecasting (NCMRWF). The Dehradun-based Forest Research Institute (FRI) is also actively engaged in the project to analyze the vegetation and microclimate and the impacts of forest fire on them.

India's forests are currently governed by the National Forest Policy of 1988, an update to which has been in the offing for nearly four years now. Key policies regarding forests and forest management are either missing, delayed or left open-ended as of date. For example, at present, there is no clear definition of forest that is accepted nationally and states are left to determine their definition of forests.

There is no central policy on inviolate forest – forests that are to be left untouched. In terms of forest management, there are regular attempts to involve the private sector. But this has experts concerned about the tribal community and the forest dwellers who could be negatively impacted. One of the constant criticisms of the central government over the years is its speedy pace of giving clearances to projects requiring forest land diversion while it drags its feet on recognising the rights of tribal people and forest dwellers over forests.

The central government had attempted to make progress on the forest rights front in 2019, when it unveiled a zero draft of an amendment to the Indian Forest Act 1927. The draft came under heavy criticism from civil society on grounds that it gives more power to the forest department authorities, including giving them the power to shoot people, and undermines the role of forest dwellers. The zero draft was withdrawn by the government in November 2019.

In the absence of an updated national forest policy, environmentalists feel, everything is being run on an ad-hoc basis.

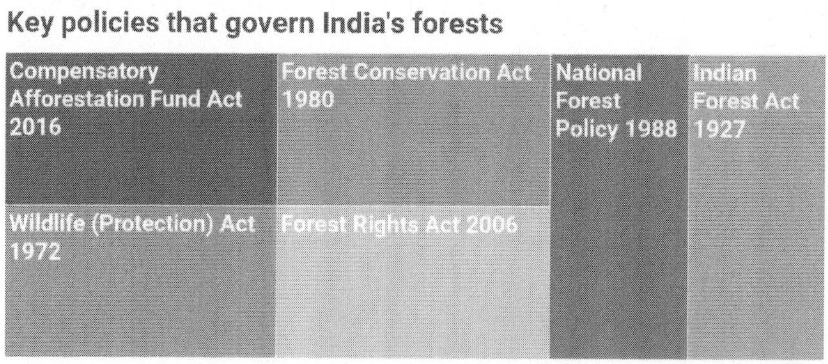

The latest India State of Forest Report 2019 which was released by the MoEFCC in 2019 emphasised that even though India's overall forest and tree cover increased in 2019 compared to 2017, the forests under the category "recorded forest area" – land notified as forest by the government – in tribal districts, which are home to about 60% of India's forests, is decreasing.

At present, nearly 25% of India's total geographical forest area is under forest and tree cover.

Compensatory afforestation

Attempts towards the dilution of forest regulations have been made with various governments. However, post 2014, with "ease of business" at the forefront of the government's action plan, dilution of forest laws gained pace as the prime minister promised to speed up clearances for industries looking at projects in forest land.

In India, the diversion of forest land for non-forestry use is governed under the Forest Conservation Act 1980, under which any organisation seeking diversion of forest area needs to support "compensatory afforestation", wherein trees are planted – either directly or via the forest department – to compensate for the lost forest.

Forest Policy for State Forest Management in India

Historically, Indian forest policies have alienated people from the forests, thereby, exacerbating the rates of deforestation. Post-independence forest policies contributed to an expansion in agricultural production, met industrial demand for raw materials, and tightened control of forest lands through restricted access to forests and forest products. Protection policies increased the hardships of vulnerable social groups by denying them access to forests. While the state took responsibility for managing forest resources, it did not have the commensurate resources to effectively manage and police the forests from traditional users. Before state intervention, forests were managed as communal property; the crucial role of forests in the economic subsistence of individuals, families and community was the basis for managing them as communal resources. A failure to recognize community control of forests led to a collapse in institutional norms that were instrumental in protecting and managing forest resources for local use. A shift in property rights to the state steadily undermined the rights of tribals to use and extract forest resources.

Involvement of rural communities living close to forests in protection and management of forest resources is enshrined in the National Forest Policy 1988.Translation of policy found expression in the resolution of Government of India, Ministry of Environment and Forests issued in June 1990.It envisaged that in lieu of the participation, the local communities will be entitled to sharing of usufructs in a manner specified by the concerned State Forest Departments. This led to the initiation of Joint Forest Management (JFM) programme. Importance of the programme is evident from the fact that the Government of India has constituted a "JFM Network" with the Inspector General of Forests, Government of India as the Chairman.

The objectives of the network are (i) to act as a regular mechanism of consultation between various agencies engaged in JFM work in the country and (ii) To obtain constant feedback from various stakeholders on the JFM programme for proper policy formulation and suitable direction to States.

World leaders adopted the Millennium Declaration at the Millennium Summit in September 2000. The proportion of land area covered by forest globally is one of the indicators for the seventh MDG i.e to ensure environmental sustainability. In addition to quantitative, time-bound targets, the Millennium Declaration calls for other actions, including intensified efforts for "the management, conservation and sustainable development of all types of forests", an international commitment to sustainable forest management made in 1992 at the United Nations Conference on Environment and Development (UNCED) and embodied in the Forest Principles and Chapter 11 of Agenda 21. Subsequent intergovernmental deliberations to promote progress towards sustainable forest management took place in the Intergovernmental Panel on Forests (IPF) and Intergovernmental Forum on Forests (IFF) from 1995 to 2000, and continue in the United Nations Forum on Forests (UNFF) as well as in other fora.

Forests and the Millennium Development Goals

The MDGs call for the integration of the principles of sustainable development into environmental policies. Environmental sustainability is being mainstreamed in forest policies around the world, particularly since UNCED, while the integration of the goals of poverty and hunger reduction in forest policies and plans is less widespread.

Community-based forestry, or participatory forestry, is particularly well placed to address poverty reduction. Community-based forestry is now well accepted and established in various countries in all regions, and programmes are beginning to generate financial and other benefits. Improving local peoples' rights and access to forest resources is a tool to the success of community-based forestry programmes. However, much still remains to be done to clarify and secure access rights. Many countries are working to strengthen forest governance, some through decentralization processes that allow the poor to derive more benefits from forests and be more involved in decision-making and forest management itself.

Although improving rights and access to forest resources and developing small-holder forest-based enterprises (including through community-private sector partnerships) show particular promise for poverty reduction, local political and economic realities, opportunity costs for the use of local resources, and other factors may prevent the poor from benefiting from community-based forestry programmes to the extent intended.

Intersectoral coordination is important for the achievement of all MDGs, but is particularly critical for reducing poverty and hunger and ensuring environmental sustainability, which are highly cross-sectoral by nature. Improved intersectoral cooperation and coordination will help efforts both to integrate the principles of sustainable development into forest-related policies and to integrate forests into sustainable development plans.

Joint Forest Management

Forest-based poverty reduction efforts tend to be linked to other land uses and should form a part of rural development strategies. Conversely, the potential for forests and trees outside forests to contribute to environmental sustainability cannot be fully realized without intersectoral cooperation and coordination. Intersectoral coordination, although difficult and time consuming, is necessary for sound decisions on land use and resource allocation, particularly when there are trade-offs between national development goals.

National Forest Policy in India treats forests as environmental and social resource. With the initiative of assigning ownership of Non Timber forest Produce (NTFP) to the local communities including the grass root level democratic institution for enhancing their livelihood opportunities and also improving their income with the value addition.

India has shifted the approach of forest management from regulatory to participatory mode of management with the resolution promulgated in 1990. At present, more than 17 million forests is managed by almost 10,000 Joint Forest Management Committees with the benefit sharing mechanism. In addition, the Government of India is in process to frame legislation for the settlement of tenurial rights of the forest dwelling communities mainly tribal on forests. This would definitely help in reducing the poverty of forest dwelling communities.

The JFM resolution was circulated by Ministry of Environment and Forest in the year 1990 and 2000. JFM is a government resolution. A government resolution is a executive order or opinion of the legislature. A resolution does not have any legal backing.

JFM as the term indicates is the management of forest by more than one party. In India there are two parties: the government represented by the Forest Department and the people living in villages located within forest or on the fringes.

There are two major reasons behind introducing JFM: one that the government's management system was not succeeding in arresting growth of forest degradation and deforestation. Second a new management paradigm was evolving in which the local people's participation was found to be an appropriate and promising tool in arresting forest degradation.

However in pre-independent India the concept of JFM didn't exist. The first National Forest Policy was adopted in 1894. Following were the guide lines:

- Ensure maintaneance of adequate forest cover
- Meet the needs of local people.
- Collect maximum revenue after meeting the needs of the local people.
- Give priority to permanent cultivation over forestry land

In post independent India there was a shift in policy. In1988, the new forest policy was adopted which covered all the sustainable management approaches. The new policy had a few unique features. Which were as follows;

- Maintenance of environmental stability and restoration of ecological balance, soil and water conservation
- Conservation of natural heritage and genetic resources.
- Increasing productivity to meet the local needs then the national need
- Creating massive peoples participation movement to protect forest and tree cover and achieve the objective of reducing pressure on existing forests and meeting peoples need.
- Deriving economic benefits must be subordinated to these principal aims.

This initiated a process of reform at the local policy and operational level of forest management ensuring that the Forest Department developed close collaboration for protection and sustainable management of forests.

The aim was –Involvement of village communities and voluntary agencies of degraded forest land. Important guidelines were as follows

- The program should be implemented under an arrangement between a voluntary agency or beneficiaries and the State Department.
- No ownership rights or lease should be given over the forest land.
- The beneficiaries should be entitled to share usufructs to the extent and subject to conditions prescribed by the State Government.
- Access to forest land usufructs should be available only to benefactress who get organized into a village institution especially for forest regeneration and protection. This could be through a village panchayat or a Village Forest Committee.
- The beneficiaries should be given usufructs like grass, lops and tops of branches and minor forest produce. If they successfully protect the forest they will be a portion from the sale proceeds when they mature.
- Areas selected from the program should be free from claims from any person who is both a beneficiary under the scheme.
- The selected site should be worked in accordance of Working Scheme duly approved by the state government .Such a scheme may remain in operation for ten years and revised after that. The working scheme is prepared in consultation of with the beneficiaries
- It should ensure that there is no grazing at al on the forest land protected by the Village Forest Committee. Permission to cut and carry grass free of cost should be given so that stall feeding is promoted.
- No agriculture should be promoted on the forest land. # Cutting of tress should not be permitted before they are ripe for harvesting. The Forest Department should not be permitted to cut to cut trees protected by the Village Forest Committee except in a manner prescribed in the working scheme. In case of emergency needs the village community should be taken into confidence.
- The Forest Department should closely supervise the work .If

beneficiaries are unable to perform their assigned duties in a satisfactory manner the usufctory benefits will be withdrawn without giving any compensation.

Such set-up, however suffered from certain flaws:

- Bye- laws have not been formulated for the functioning of the JFMC though now most of the states have issued executive orders for the functioning of JFMC but these executive orders are not binding on JFMC. The aim was to decentralize the process and make JFMC.
- Minor Forest Produce (MFP) has not been defined neither by the State legislature or by the Centre.
- Central and State/UT Governments have issued guidelines for the creation and functioning of JFMC but these guidelines are not in conformity with the provisions of the Constitution
- Through the Constitution (Seventy-third Amendment) Act, 1992 and Panchayat (Extension to the Scheduled Areas) Act, 1996 ownership rights over minor forest produce (MFP) have been given to Village Panchayats. Now on the same resource base i.e. NTFP/MFP we have two sets of groups having ownership and while the Panchayats have a legal backing, JFMC don't have. conservation have been raised while the proponents of Panchayats have termed JFM as a parallel institution against the spirit of the Constitution.
- The JFM resolution only provides 20% of share to be given the Joint Forest Management committees , while rest of the income would go to the Forest Department. There is an obvious unequal distribution of benefit sharing between the parties. These committees would have to protect the forest for ten years ,and would only receive 20% of the share.
- The national resolution provides that at least 33% of the seats shall be reserved for women in the specified committees. However states like Rajasthan are not following this provision.

Panchayati Raj System

The concept regarding Panchayati is contained in the Part IX of the 73rd amendment of The Indian Constitution .This amendment came into force in on 24th April 1993. It institutionalized the third stratum of government (Panchayats) at the local level.

Through the Constitution (Seventy-third Amendment) Act, 1992)

Part IX "The Panchayats" was inserted in the Constitution which paved the way for "Village Panchayats" by making provisions for the constitution of Panchayats, their composition, election, powers, authority, responsibility, audit, etc. This Act doesn't apply to Schedule Area referred to in clause (1) and tribal areas referred to in clause (2) of Article 244 and certain other specified areas Village Panchayats have been given the responsibility of social/farm forestry, minor forest produce (MFP), and soil conservation through Eleventh Schedule.

"Panchayat" means an institution (by whatever name called) of self-government constituted under article 243B, for the rural areas. In the section 243B of the amendment it fails to define what is 'self government'. It does not clarify whether self government means complete autonomously or extension of the state .It is left to states to derive their own interpretation regarding the nature of self government at the local level .By the word "nature" of self government it is meant reservation to be provided for the marginalized section of the society, the powers given to each level of self governing relating to administration, execution and performing the defined roles would be different in each State . Thus this will defeat the purpose of uniformity regarding the nature and concept of self government at an all India level.

The flaw again lies that the amendment has failed to assert the nature of 'Power And Authority' .This task is left to the state legislatures .Again defeating the purpose of uniformity. The consequences will be such that either the states will deny power to the panchyats instead of becoming autonomous local governing units they will basically remain in bureaucratic control .The burecratic trend can be seen most of the states .It is seen in the State of Goa power and functions have not yet been transferred to the Panchayats.

As per the law there should be continuous elections every five years .However the statistics says that this status is not achieved in most of the states .None of the states except WestBengal, Tripura and Rajasthan have not been able to hold continuous elections every five years. The judgment provided by the Supreme Court states that it is mandatory for every state to hold elections to Panchayat. Giving powers to the third stratum of government for effective management of resources at the district ,intermediate and village level. Parallel institutions have come about for management of the natural resources which.

Participatory Approach

According to United Nations Development Programme's report 67.7

million people belonging to "Scheduled Tribes" in India are generally considered to be 'Adivasis', literally meaning 'indigenous people' or 'original inhabitants', though the term 'Scheduled Tribes' is an administrative term used for purposes of 'administering' certain specific constitutional privileges, protection and benefits for specific sections of peoples considered historically disadvantaged and 'backward'. Out of the 5653 distinct communities in India, 635 are considered to be 'tribes' or 'Adivasis'. With the ST population making up 8.08% (as of 1991) of the total population of India, it is the nation with the highest concentration of 'indigenous peoples' in the world. 68 million tribals who inhabit forests and wild lands throughout India, Adivasis have evolved an intricate convivial-custodial mode of living.

The Beginning of Land alienation

Introduction of the alien concept of private property began with the Permanent Settlement of the British in 1793 and the establishment of the "Zamindari" system that conferred control over vast territories, including Adivasi territories, to designated feudal lords for the purpose of revenue collection by the British. This drastically commenced the forced restructuring of the relationship of Adivasis to their territories as well as the powe relationship between Adivasis and 'others'. The predominant external caste-based religion sanctioned and practiced a rigid and highly discriminatory hierarchical ordering with a strong cultural mooring.

After the transfer of power, the rulers of the Residency Areas signed the "Deed of Accession" on behalf of the ruled on exchange they were offered privy purse. No deed was however signed with most of the independent Adivasi states. They were assumed to have joined the Union. The government rode rough shod on independent Adivasi nations and they were merged with the Indian Union. This happened even by means of state violence as in the case of Adivasi uprising in the Nizam's State of Hyderabad and Nagalim.

The Constitution of India, which came into existence on 26 January 1950, prohibits discrimination on grounds of religion, race, caste, sex or place of birth (Article 15) and it provides the right to equality (Article 14), to freedom of religion (Articles 25-28) and to culture and education (Articles 29-30). STs are supposedly addressed by as many as 209 Articles and 2 special schedules of the Constitution - Articles and special schedules which are protective and paternalistic.

Article 341 and 342 provides for classification of Scheduled Castes (the untouchable lower castes) and STs, while Articles 330, 332 and

334 provides for reservation of seats in Parliament and Assemblies. For purposes of specific focus on the development of STs, the government has adopted a package of programmes, which is administered in specific geographical areas with considerable ST population, and it covers 69% of the tribal population.

Forest Laws in Relation to Tribal Land Conflict

The total forest cover in India is reported to be 765.21 thousand sq. km of which 71% are Adivasi areas. Out of this 416.52 and 223.30 thousand sq. kms. is categorised as reserved and protected forest respectively. About 23% of these are further declared as Wild Life Sanctuaries and National Parks which alone has displaced some half a million Adivasis. By the process of colonisation of the forests that began formally with the Forest Act of 1864 and finally the Indian Forest Act of 1927, the rights of Adivasis were reduced to mere privileges conferred by the state.

The Imperial Forest Department was formed in 1864. The first Act for the regulation of forests was passed in 1865. It empowered the government to declare any land covered with trees or brushwood as government forest and to make rules to manage them. The act was applicable only to the forests in control of the government and did not cover private forests. It made no provision regarding the rights of the users. Its primary function was to generate revenue by selling timber and forest produce. Since forests are deemed to be government property, Adivasis were barred from collecting forest produce.

According to the Indian Forest Act, 1927, the Government can constitute any forest land or waste land which is the property of Government or over which the Government has proprietary rights, a reserved forest, by issuing a notification to this effect. Commercial interests of the then British Government motivated it to declare more and more lands as reserved forests, without ascertaining the rights of the tribals and other forest dwellers.

After the enforcement of the Indian Forest Act in 1927, the government declared all lands not claimed by private individuals and agencies as forest lands and classified them into reserve, protected and village forests. To attract labour to work in forestry related activities, the governments gave lands to households for cultivation and also promised jobs. A number of forest villages were established. The lands allotted for cultivation were on a purely temporary basis and forest officials were to look after the administration of these villages.

After Independence, the inhabitants of forest villages realized that most villages lacked infrastructure facilities; they could not get loans for development of agriculture as they had no title over the land. Around 1980, the government decided to convert forest villages into revenue villages. This was done in many states. However, not all the forest villages have been converted into revenue villages. In some villages, land has been given on a 15 year lease to enable the holders to obtain loans from banks. Residents have agitated for the conversion of these villages into revenue villages.

Even after independence in 1947, during the process of amalgamation of princely states, the activity of consolidation of government forests continued. The State Governments / UT Administrations proclaimed the lands of ex-princely states and the zamindar-lands as Reserved Forests. However, no effective steps were taken to give an alternative accommodation to the tribal people residing in this area. Bar on accrual rights are imposed through section 5 of the Indian Forest Act.

The Act invites claims for settlement in the reserve forest area. This is provided from section 3 to section 20 of the IFA. These sections provide a complicated procedure which requires documentation in relation to ownership of the land, evidence, which would be impossible for an indigenous to produce in front of the Settlement Officer. Many tribal families today are regarded as "encroachers" on forest land (persons who are cultivating or residing on forest land without a legal right to be there).

In many cases, however, it would be more accurate to say that the "forest" is encroaching on their traditional rights. There are a number of core issues that are central to the resolution of forest based conflicts, and are integral to the question of encroachment.

Protected Areas

About 4.5% of the total land in the country is covered under protected areas, classified into national parks, wildlife sanctuaries and protected areas. The Wildlife (Protection) Act of 1972 provided for the administration of these areas. The act places several restrictions on the residents of villages in these areas. Efforts are also being made to relocate these villages outside such areas. These efforts have met with stiff resistance from the residents and violent conflicts have taken place in many places. Under pressure from the World Bank and the environment protection fund agencies, the government has tightened the restrictions on the hunting of animals included in the list of wildlife species, as also a number of economic activities. Consequently, development projects in these areas have been shelved.

Throughout this, one needs to be acutely aware that there are huge forest encroachments by powerful vested interests which need to be dealt with strictly. Mechanisms will need to be put in place to ensure that powerful interests/ land mafias do not use poor people as a front to further their encroachments. In the past two decades even as the genuine forest dwellers have lived in the fear of eviction, mafias have continued to encroach forests and alienate tribal land. This has not only had severe ecological implications due to disturbance of forest and wildlife but also displaced tribals from their homes and livelihoods.

Forests Conservation Act

The 42nd Constitutional Amendment shifts forests from the "State List" to the "Concurrent List". The FCA prohibits non-forest use of forest land without the government's approval. It also advocates "sustainable forest management through participatory approach", with "due regard to the traditional rights of the tribal people on forest land". Further making the claim settlement process stringent and inflexible and thereby the settlement dispute continues for decade.

The 1990 guidelines on Tribal and Forest Interface expressly provides for speedy settlement of claims in relation to forest land, Regularization of Encroachments; Review of Disputed Claims over Forest Land; Regularization of Pattas & Leases; and Conversion of Forest Villages to Revenue Villages. These Guidelines were supposed to provide a framework to resolve the problem of settlement of rights of tribals and other forest dwellers on forest land. For instance, FP1 outlined procedures whereby state governments could apply to the Government of India for regularization of "pre-1980 encroachments". As discussed below, however, these Guidelines have not been implemented.

On 5 February 2004, the MoEF issued new guidelines for "regularization of the rights of the tribals on the forest lands", in continuation of the 1990 Guidelines. These "supplementary guidelines" request the State Governments to give legal recognition to "the traditional rights of the tribal population on forest lands", and to submit proposals for conferring "heritable but inalienable rights over such lands" on "tribal dwellers who are in continuous occupation of such forest land at least since 31/12/1993".

There are thousands of cases of local inhabitants claiming that they were in occupation of notified forestlands prior to initiation of forest settlements under the Indian Forest Act. There are a number of cases of pattas/leases/grants said to be issued under proper authority but which has now become contentious issues between different departments, particularly the Forest Department and the Revenue Department. The problem is compounded by the fact that in many cases there is no clear demarcation of forest lands. In fact most of the disputes and claims relating to use and access to forests have lingered on and evaded resolution in the past because of the failure to demarcate precisely the extent of the forest. All of these require remedies and an approach aimed at only evicting the forest-dwellers is worsening the situation, not remedying it.

There is a need to consciously distinguish between the authorised/ unauthorized occupation of forests by dwellers for bonafide consumption, and the use of forest resources from the large scale timber and fuel wood smuggling and intensive commercial exploitation of forest resources. In the former case the so-called encroachment is a local and subsistence-oriented activity, while in the latter case it is a widespread organized industry largely driven by the mafia. The approach of the law and policy to the two situations cannot be the same

The National Forest Policy was adopted in 1894. Following were the guide lines:

- Ensure maintenance of adequate forest cover
- Meet the needs of local people.
- Collect maximum revenue after meeting the needs of the local people.
- Give priority to permanent cultivation over forestry land

In post independent India there was a shift in policy. The National Forest Policy Resolution1952 adopted by the government, emphasised that the forest policy should be based on paramount national needs. For the first time the resolution emphasized the ecological and social aspects of forest management, giving secondary importance to the needs of commerce, industry and revenue. The resolution did not call for any change in the forest law and remained only a pious declaration.

The National Forest Policy, 1988, has a separate section on 'Tribal people and forests' which states that: "having regard to the symbiotic relationship between the tribal people and forests, a primary task of all agencies responsible for forest management, including the forest development corporations should be to associate the tribal people closely in the protection, regeneration and development of forests as well as to provide gainful employment to people living in and around the forest." The section on 'rights and concessions' has focused on 'full protection' of the rights of tribals.

In1988, the new forest policy was adopted which covered all the sustainable management approaches. The new policy had a few unique features. Which were as follows;

- Maintenance of environmental stability and restoration of ecological balance, soil and water conservation
- Conservation of natural heritage and genetic resources.
- Increasing productivity to meet the local needs then the national need
- Creating massive peoples participation movement to protect forest and tree cover and achieve the objective of reducing pressure on existing forests and meeting peoples need.
- Deriving economic benefits must be subordinated to these principal aims.

This initiated a process of reform at the local policy and operational level of forest management ensuring that the Forest Department developed close collaboration for protection and sustainable management of forests.. While the resolution adopted a pro-tribal policy, the old Act of 1927 with all the subsequent amendments remained unchanged.

The Scheduled Tribes and Other Traditional Forest Dwellers (Recognition of Forest Rights) Act, 2007

The recently notified Act aims to compensate the "historical injustice" done to forest-dwelling tribes that were denied their traditional rights to forest lands and resources in the last couple of hundred years. In this period, the lands they had been dwelling on were declared "forest land" or "protected areas" for wildlife. The traditional conservation ethos and dependence of tribal societies on forest land were overlooked. The Act aims to reverse the alienation of tribes from their own habitat caused by past policies and laws.

The rights proposed to be accorded to Adivasis include pattas to forest lands occupied before 1980. This is not new; it has been a stated policy of the government for years now. A number of other rights are also to be granted: nistar (usufruct) or ownership rights to forest resources, grazing rights including seasonal ones of nomadic communities, habitation rights (for those classified as Primitive Tribal Groups), conversion of forest villages into revenue villages, and so on.

No tribal person is to be evicted from currently occupied land until the process of determining rights is completed.

Under the Act

- All rights are accompanied by responsibilities for forest protection;
- All right holders have the duty to conserve forests and wildlife, protect catchment areas, water sources, and ecologically sensitive areas, and inform the gram sabha of ecologically destructive activities;
- Hunting is explicitly excluded from the list of forest rights;
- All rights are meant only for bona fide livelihood needs, not for exclusive commercial purposes;
- In no case would forest land beyond 2.5 hectares be allotted (only land under occupation prior to 1980 can be considered, no new forest lands would be allotted).

The MOEF is of the view, land settlement of the forest dwellers would increase the problem of encroachment to the already depleting forest cover in our country, poaching would be rampant.

Case Laws

One of the earliest cases relating to forest rights was that of Banwasi Seva Asharam v. State of Uttar Pradesh . The case was initiated as a public interest writ petition under article 32 of the Constitution on behalf of the local people protesting for reservation of forest land by the state.

People in the village have lived for generations and had relied for their sustenance on the forest products. The petitioners had alleged that state has ignored their claims of these people over the forest and steps were being taken foe eviction of many forest dwellers. A high level committee was formed by the apex court to look into the matter of claims.

In the meantime, the government informed the court it wanted to site National Thermal Power Station over the disputed forest land. The land was granted by the by the court to NTPC, thereby ordering eviction of the forest dwellers.

The reasoning given by the court that forests are as much national assets, however generation of electricity would be equally important. The court did not address the primary aim of claim of the forest dwellers arising out of their dependence on the forest life guaranteed by article 21 of the Indian Constitution.

The court dropped the all the claims over the land which the state has notified as a reserved forest area.

The claims of the petitioner were written off on procedural grounds. The written application were not stamped, oral objections would not be considered, the tribal were ignorant of their legal rights.

The NTPC asked to rehabilitee the tribal by housing transportation employment, infrastructure.

In **T.N Godavarman Thirumulkpad v. Union of India**, the apex court issued directives sweeping directives to oversee the enforcement of forest laws across the nation. The court froze all forest wood based industries, enforced the scope of embargo on forest exploitation, and issued directions for sustainable forest management. The court focused the issues affection North East region where wood based industry contributes sustainability of the regions economy. The issue raised by the amicus curiae was whose interest should be protected, the forest dweller or forest persons whose livelihood depends upon.

The court stated that in depth examination of the Forest policy should be taken care off.

It should conform to the directions of National Forest Policy 1988 As stated in the policy the respective department of the states should encourage tribal welfare and joint forest management.

International Perspective

The right of indigenous peoples have been recognized by international agencies and organizations. Every individual has right to have a living, proper housing, drinking water, education, electricity, health facilities, communication, right over the land, they are denied everything.

United Declaration of Human Rights(1948) and International Covenant on Civil and Political Rights (1966) have specific significance for indigenous peoples. The Universal Declaration provides a common standard for the human rights of all peoples and all nations, and proclaims the importance of traditional, political, and civil rights, as well as basic economic social and cultural rights e]veryone has the right to own property alone as well as in association with others."

Furthermore, "[n]o one shall be arbitrarily deprived of his property

. The Covenant spells out civil and political rights and guiding principles based on the Universal Declaration.

The 1957 International Labour Organisation Convention , Protection and Integration of Indigenous and Other Tribal and Semi-Tribal Populations in Independent Countries, addresses the right of indigenous peoples to pursue material well-being and spiritual development, and was a first international instrument in specific support of indigenous peoples. Largely because of its view that indigenous peoples should be integrated into the larger society, a view that subsequently came to be seen by many as inappropriate.

It was followed by Convention Concerning Indigenous and Tribal Peoples in Independent Countries. It presents the fundamental concept that the way of life of indigenous and tribal peoples should and will survive, as well as the view that indigenous and tribal peoples and their traditional organizations should be closely involved in the planning and implementation of development projects that affect them . The convention specifically provides recognition of traditional land rights of the indigenous people . The concentration gives the indigenous people the right to sustain themselves through natural resources, the collection of forest produce and speedy recovery of traditional rights of the indigenous people.

India has signed convention on Protection and Integration of Indigenous and Other Tribal and Semi-Tribal Populations in Independent Countries, but not Convention Concerning Indigenous and Tribal Peoples in Independent Countries.

The 1992 Convention on Biodiversity calls on contracting parties to respect traditional indigenous knowledge with regard to the preservation of biodiversity and its sustainable use. The Vienna Declaration and Programme of Action emerging from the 1993 World Conference on Human Rights recognizes the dignity and unique cultural contributions of indigenous peoples, and strongly reaffirms the commitment of the international community to the economic, social, and cultural well-being of indigenous peoples and their enjoyment of the fruits of sustainable development.

The draft statement of United Nations Draft Declaration on Rights of Indigenous People, expressly states that traditional lands of the indigenous people should be returned, if not they should be equally be compensated. The United Nation Forum on Forest in. the fifth session expressly stated sustainable development and management of forest would not be possible without the participation of indigenous community residing in the forest. India though a signatory to many of the convention stated above, yet has failed to implement the provisions stated in hem through relevant legislation.

Conclusion

Planning Commission estimates suggest that 21.3 million people were displaced by development projects between 1951 and 1990 alone. Of these, 8.54 million (40%) belonged to Scheduled Tribes constituting only 8% of the total population. Only 2.1 million (25%) of them are reported to have been rehabilitated. The rest were left to fend for themselves. Cases fixing of responsibility on project proponents is important and a mechansim needs to be developed to provide reparations and complete and just rehabilitation before passing 'eviction' . In the absence of such a process such people will be displaced unfairly once again.

It is important that any proposal for dealing with such encroachments needs to first ensure stoppage of tribal land alienation and restoration of alienated land to these tribals. In fact land restoration in such cases needs to be a mandatory first requirement. The nistar rights of the forest dwellers should be recognized in relation to minor forest produce for sustenance.

The Indian Forest Act,1927 and the Forest Conservation Act,1980 should be amended in relation to National Forest Policy 1988 and the Joint Forest Management resolution. Furthers the Indian Forest Act 1927 and the Conservation Act 1980, need to be amended in relation to the Forest Policy and the Joint Forest Management Resolution. Unless and until the Act and policy serve the same purpose their will always be a dispute in relation to the rights and duties. It is regarded that forest are national asset and in the present era, sustainable development is the key word. However this does not mean that rights of the tribals will be denied to them. It is a known fact that Sustainable development can only be achieved through participatory approach of the government with the indigenous people.

Total Forest and Tree Cover rises to 24.56 percent

Compared to the assessment of 2017, there is an increase of 5,188 sq. km in the total forest and tree cover of the country. "Out of this, the increase in the forest cover has been observed as 3,976 sq km and that in tree cover is 1,212 sq. km; Range increase in forest cover has been observed in open forest followed by very dense forest and moderately dense forest and the top three states showing increase in forest cover are Karnataka (1,025 sq. km) followed by Andhra Pradesh (990 sq km) and Kerala (823 sq km).

Some Major Findings

Area-wise Madhya Pradesh has the largest forest cover in the country followed by Arunachal Pradesh, Chhattisgarh, Odisha and Maharashtra. In terms of forest cover as percentage of total geographical area, the top five States are Mizoram (85.41%), Arunachal Pradesh (79.63%), Meghalaya (76.33%), Manipur (75.46%) and Nagaland (75.31%).

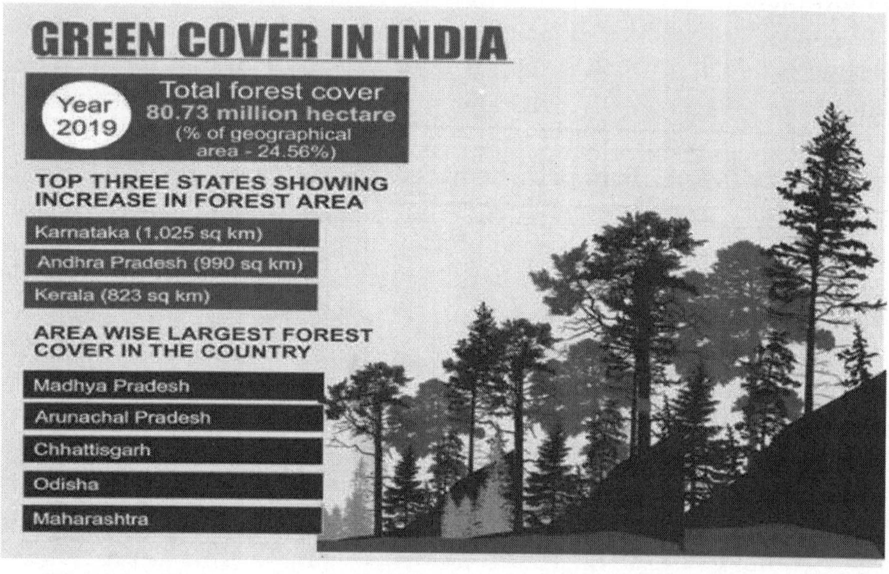

The Mangrove ecosystems are unique & rich in biodiversity and they provide numerous ecological services. Mangrove cover has been separately reported in the ISFR 2019 and the total mangrove cover in the country is 4,975 sq km. An increase of 54 sq Km in mangrove cover has been observed as compared to the previous assessment of 2017. Top three states showing mangrove cover increase are Gujarat (37 sq km) followed by Maharashtra (16 sq km) and Odisha (8 sq km).

The total growing stock of India's forest and TOF is estimated 5,915.76 million cum of which 4,273.47 million cum is inside the forests and 1,642.29 million cum outside. There is an increase of 93.38 million cum of total growing stock, as compared to the previous assessment. Out of this the increase in growing stock, there is an increase of 55.08

million cum inside the forests and 38.30 million cum outside the forest areas.

The extent of bamboo bearing area of the country has been estimated 16.00 million hectare. There is an increase of 0.32 million hectare in bamboo bearing area as compared to the last assessment of ISFR 2017. The total estimated green weight of bamboo culms is 278 million tonnes, slowly an increase of 88 million tonnes as compared to ISFR 2017.

Under the current assessment the total carbon stock in country's forest is estimated 7,124.6 million tonnes and there an increase of 42.6 million tonnes in the carbon stock of country as compared to the last assessment of 2017. The annual increase in the carbon stock is 21.3 million tonnes, which is 78.2 million tonnes CO2 eq.

Wetlands within forest areas form important ecosystems and add richness to the biodiversity in forest areas, both of faunal and floral species. Due to importance of wetlands, FSI has carried out an exercise at the national level to identify wetlands of more than 1 ha within RFA. There are 62,466 wetlands covering 3.8% of the area within the RFA/GW of the country.

Need for rethinking management strategies

Earth's landscape has been subjected to both natural and anthropogenic fires for millions of years.

Natural, lightning-caused fires are known to have occurred in geological time continuously at least since the late Silurian epoch, 400 million years ago, and have shaped the evolution of plant communities.

Hominids have used controlled fire as a tool to transform the landscape since about 700,000 years ago. These hominids were *Homo erectus*, ancestors of modern humans. Paleofire scientists, biogeographers and anthropologists all agree that hominid use of fire for various purposes has extensively transformed the vegetation of Earth over this period.

The nature of Earth's modern-day biomes would be substantially different if there had been no fires at all. William Bond and colleagues (2005) used a Dynamic Global Vegetation Model to simulate the area under closed forest with and without fire. They estimated that in the absence of fire, the area of closed forest would double from the present 27% to 56% of present vegetated area, with corresponding increase in biomass and carbon stocks. This would be at the expense of C4 grasslands and certain types of shrub-land in cooler climates.

Figure : Dry season ground fire in Mudumalai.

Native peoples have extensively used fire in recent millennia to shape their habitats. For instance, Native Americans used fire throughout the continent to increase the production of berries, seeds, nuts and other gathered foods for their subsistence (Krech 1999 – The Ecological Indian). In Australia, Garry Cooke and colleagues (2011) have similarly documented the complex Indigenous fire management practices of aboriginal tribes whose history of migration to the continent goes back about 50,000 years before present. In both cases, the arrival of Europeans resulted in a vastly different outlook on land management. Fire suppression policies began to be enforced.

In India, too, the tribal people traditionally used fire as an essential part of land (including forest management). European foresters trained in silvicultural practices that saw fire as a hindrance to tree growth began enforcing fire suppression policies in Indian forests. Fire suppression went beyond timber production – to quote Stephen Pyne (1994), "to control fire was to control native populations". "All fire is bad" eventually became an integral part of India's conservation discourse and continued into the period since Indian Independence.

Dry forests cover a major part of India's land surface. Given the dense human population and various anthropogenic activities, it is extremely difficult to keep fire out of forests for any reasonable length of time. Lightning-caused fires also happen in Indian forests, especially in the north, though the extent and frequency are not known.

Fires are most destructive when they occur late in the dry season, under conditions of high temperature and low humidity. The universal fire suppression policy of the forest department in all forest types (the exceptions being grassland of the floodplains) results in a build up of fuel load (litter, grass and woody debris) over the years. Fires are successfully prevented for several years increasing the fuel load. Then, a combination of low humidity and high temperatures late in the dry season creates the perfect conditions for a very high intensity and destructive fire when a deliberate or careless action by people sets the ground vegetation on a high temperature fire with the potential to cause a canopy fire.

There is increasing scientific evidence that the invasion of woody plants such as *Lantana camara* in southern Indian forests is the result of fire suppression over the years with consequent high intensity fires.

Figure: Lantana flowers.

Should we control fires? YES.

However, a pragmatic fire management policy for dry forests should look at the option of learning from indigenous fire management practises such as early dry season, cool burning to avoid late dry season high temperature fires.

Much research is needed on fire management.

We should not be afraid of using fire judiciously as tool to manage forest fires to achieve broader goals of increasing forest diversity, biomass and carbon stocks. Mere fire suppression will not serve the purpose as the experience with many developed countries (USA-California, Portugal, Spain, Greece, Australia) has shown in recent years (all these have experienced highly destructive fires in recent years).

Tropical moist forests such as rainforests should, of course, be strictly protected against fires.

Forest fire risk mapping using analytical hierarchy process (AHP) and earth observation datasets

Natural conservation areas that play an important role in maintaining the environmental balance are found in developing tropical regions forests. Despite being of paramount importance, forest cover loss in these regions is occurring at an alarming rate. Over the last few decades, decline in forest cover has been attributed to wildfires, rapid economic development, agriculture, logging, and increasing human population. Globally, approximately, 30% of tropical forests were degraded by logging or fire between 2000 and 2012.

In India, approximately, 64% of the total forest area is frequently affected by forest fires , causing an economic loss of over 440 crores . Forest fires can be caused naturally (e.g., lighting) and/or induced via human activities, however, in India the latter is dominant. About 90% of the forest fires in India are caused by humans . Over the last few years, forest fires in India have received greater attention because of their ecological, economic, social, climatic, and political impacts . Despite the fact that forest fires are mostly regulated by anthropogenic factors, local environment, metrological condition, and topography factors also play an important role in their uncontrolled expansion. In this regard, it is very important to identify vulnerable areas that come under “forest fire risk zones.” Therefore, parameters that play an important role in forest fire need to be quantified timely with advanced technology.

Timely and accurate information of complex fire models needs spatial and temporal information that is furnished by geospatial technologies. Additionally, integration of multi-criteria decision-making (MCDM) in spatial domain provides a novel framework toward the assessment of fire risk zones. The AHP method is among the most widely used MCDM . AHP along with geospatial techniques has been successfully utilized for identifying the fire risk zones in many parts of the world.

Assignment of weight to different thematic layer using AHP

Assessment of weight of each thematic layer is the most important for prediction of forest fire zone as output is mostly dependent on appropriate weight. Most important methodologies were adopted for calculating weightage to delineate the fire risk zone such as fuzzy sets , linguistic variables , and AHP . However, AHP is the most frequently used method for determining fire risk zone. AHP was a decision added method that generated relative ratio scales of paired comparison developed by Saaty . In the present study, AHP-based pair-wise comparison matrixes were made between different thematic layers at each level of the hierarchy.

A pair-wise comparison matrix was made to compare all factors against each other based on their importance (equal, moderate, strong, very strong, and extremely strong). A standard Saaty's 1–9 scale was used to determine the relative importance values for all themes and their respective features, where value '1' denotes "equal importance" between the two themes, and the value '9' denotes the "extreme importance" of one theme compared to the other one. The normalized relative weight was calculated by dividing each element of the pair-wise matrix by the sum of its column. The weights of each layer were decided using the principal eigenvector of the square matrix of each criterion. The higher was the weights the more was the influence of the parameters on the forest fire based on its relative importance.

CHAPTER 6

DETECTION, PLANNING, ORGANIZATION AND TECHNIQUES OF FOREST FIRE CONTROL

Introduction

Forests play a vital role in maintaining the Earth's ecological balance. Unfortunately, the forest fire is usually only observed when it has already spread over a large area, making its control and stoppage arduous and even impossible at times. The result is irreparable damage to the atmosphere and environment, where 30% of CO_2 in the atmosphere produced by forest fires. Among other consequences of forest fires are long-term disastrous effects such as impacts on local weather patterns, global warming and extinction of rare species of the flora and fauna.

Forests are vast remote abandoned areas, full of highly combustible material with dry leaves and branches to the Earth surface composites, where these are perfect to act as a fuel source for fire ignition and later fire stages. The fire ignition may be caused through human actions or by natural reasons. The initial stage of ignition is normally referred to as "surface fire" stage. This may then lead to feeding the fire flame, thus becoming "crown fire." Mostly, at this stage, the fire becomes uncontrollable, and the damage to the landscape may become excessive and could last for a very long time depending on prevailing weather conditions and the terrain.

Problem size

Forest fire is a global environmental problem causing extensive damage every year. According to International Union for Conservation of Nature (IUCN) Report "Global Review of Forest Fire 2000," wild fire is a natural phenomenon. However, over 90% of all wildland fires are due to human action causing significant forest loss, that is 6–14 million hectares of forest per annum, and 30% of the CO_2 in the atmosphere produces from forest fire. This leads to enormous economic losses, damage to environmental, recreational and amenity assets, global warming and loss of life. There is a strong recognition that action is needed to catalyse a strategic international response to forest fires .

In the case of the USA

In the 2003–2004 wildfire sieges, CAL FIRE's fire suppression costs exceeded $252.3 million; property damage costs exceeded $974 million; 5394 structures were destroyed; and more than 23 people lost their lives as a result of California wildfires. "Increasingly destructive wildfires are ravaging homes and businesses in more than three-fourths of the states. One of the most devastating fires in recent history was the $1 billion Witch Creek Wildfire that decimated vast parts of San Diego County, California, in October 2007. By the time it was fully contained, the fire had burned an estimated 198,000 acres and damaged or destroyed more than 1200 homes and 500 outbuildings."

In terms of scale, the 2007 fire season was second only to last year in acres burned and costs expended. In 2007, there were 27 fires costing over $10 million whose total suppression cost approached $547 million, exclusive of burned area emergency rehabilitation costs. These fires alone accounted for just less than 3 million burned acres. All wildfire acres reported to the National Interagency Coordination Centre in calendar year 2007 totalled 9.32 million acres at a federal cost of approximately $1.8 billion. On 17 June 2002, an estimated $9,403,000 was spent battling 196 wildland fires that scorched over 51,000 acres of land in parts of 11 states in the USA .

Hundreds of thousands of acres is burnt within the wildland urban interface (WUI) each year. Each year, over $100 million is spent on suppression efforts and more in the disaster recovery phases of catastrophic, natural and/or human-caused hazards, but the losses continue to mount.

In the case of Canada

Canada has approximately 10% of the world's forests. However, about 7400 forest fires have occurred in Canada every year over the last decade, burning an average of 1.9 million hectares of forests. In British Columbia alone in 2006, which was not a peak fire year, there were 2590 forest fires destroying 131,086 hectares and costing $156 million. Fire suppression expenses during the last decade in Canada have exceeded the $500 million and almost hit the $1 billion a year. [Facts about Wildland Fire in Canada, 2012.] On 5 October 2009, for example, a huge fire in the San Bernardino National Forest, CA, burned 3500 acres. More than 500 firefighters found it hard to control due to the strong 72 km/h wind speed . The story continued year after year in Canada.

Table 1. Forest fire in Canada.

Year	Number of fires	Number of hectares burned	Total cost (millions)
2006	2590	131,086	$156.0
2005	976	34,588	$47.2
2004	2394	220,516	$164.6
2003	2473	265,050	$371.9
2002	1783	8539	$37.5
2001	1266	9677	$53.8
2000	1539	17,673	$52.7
1999	1208	11,581	$21.1
1998	2665	76,574	$153.9
1997	1175	2960	$19.0
1996	1358	20,669	$37.1
1995	1474	48,080	$38.5

In the case of Europe

On 14 August 2012, the following report was published in ScienceDaily—'the 2012 fire season has been characterized by a high number of fires in the early season. Over 100,000 hectares had already been consumed by fire at the end of March. July brought critical fire episodes in Spain and Portugal, which led to a number of human casualties'.

There was a high fire risk in southern Europe during the last period, specifically in France, Portugal, central and southern Italy, the Balkan region, Spain, Greece and Turkey.

The European Forest Fire Information System (EFFIS) provides monitoring and fire mapping, regular forecasts for fire danger up to 6 days in advance and provides near-real time estimations for fire damages across Europe.

Until present, almost 580,000 hectares, in the area monitored by

EFFIS, have been burnt this year, which includes Europe, North Africa and Middle East countries [ScienceDaily 2012]. In August 2009, forest fires wreaked havoc in Greece (3700 acres, hundreds of houses). Also in 2007, there was a terrible fire in Greece.

In the case of Australia

On 21 November 2009, in Tasmania, Australia, a horrible forest fire was reported . Some of the fires could be natural disasters or directly or indirectly the product of human negligence and abuse of the environment (including the rise of temperature associated with the global warming).

"In February 2009, Victoria was devastated by one of the worst bushfires in Australia's history. The Black Saturday fires caused many deaths and injuries and directly affected many towns and communities, destroying homes, businesses, schools and kindergartens. Key statistics from the Victorian Bushfire Reconstruction and Recovery Authority are:

- deaths 173
- area burnt 430,000 hectares (including 51 towns, 78 communities)
- total property dollar losses $1.35 billion
- homes lost 2129, valued at $713 million (includes contents and buildings)."
- Too many horrible fires occurred in Australian history, probably that resulted from its nature.

In the case of the UK

South Wales Fire and Rescue Service cover an area of 2712 km squared, which is divided into 10 "Unitary Authorities": Bridgend, Vale of Glamorgan, Rhondda, Cynon Taff, Cardiff, Caerphilly, Merthyr, Blaenau Gwent, Torfaen, Newport and Monmouthshire, where a population of 1.4 million covered by their services .

The Welsh society statistics provided by the Fire and Rescue Services of Wales continues to be high. It shows that, in purely economic views, the expenditure to Wales's areas in years 2006 and 2007 was £56.3 million. In South Wales area alone, the cost was about £35.2 million.

Forest fire monitoring systems can help in fire suppression. The fire department will receive the information and evaluate the situation. The known rules apply here: 1 min—1 cup of water, 2 min—100 liter of water, 10 min—1000 liter of water.

Overview of forest fire behaviour

Let us define the fire first:

"Combustion is a complex process in which fuel is heated, ignites, and oxidizes rapidly, giving off heat in the process. Fire is a special case of combustion—self-perpetuating combustion characterized by the emission of heat and accompanied by flame and/or smoke. With fire, the supply of combustible fuel is controlled by heat given off during combustion"

Fire requires three components: fuel, heat and oxygen to be present. If one of the factors is missing, the fire will go out. There must be:

1. a source of fuel for combustion,
2. a source of heat to promote the reaction (the fire itself),
3. oxygen in fair concentration to maintain the reaction.

Figure : Fire triangle factors

A forest fire is a dynamic phenomenon that changes its behaviour in time. Fire spreads through forest fuel. It is performed by a complex heat transfer and thermo chemical processes that determine fire behaviour.

There is more than 27 mathematical models defined to describe the forest fire behaviour; each model was built according to different experiences in different countries with forest fire. Every model differs from the others according to the input and the environmental parameters (fuel indexing). Researchers in some countries manage to use some of these models in simulation programs or even create their own methods for mapping the landscape and the fire behaviour on monitoring screens for analysis and expectation of fire behaviour to help the firefighters to determine the best method to extinguish the fire such as BehavePlus, FlamMap, FARSITE, Geodatabase and ArcSDE. On the other hand, researcher's target is to find a reliable technology that can detect and localise the fire, help in decision-making of crisis possibility or a high fire risk situation. As a result, the fire can be extinguished in early stages within a short time to minimise the damage save lives, environment, firefighter equipment, time and effort.

A wildfire burning in constant environment takes the shape of an ellipse. Environment can be a variable of time. A fire might have different parts burning in different environments, such as moisture contents, wind speed, wind direction, slope and so forth. The environment heterogeneity might result in a very complex fire shape.

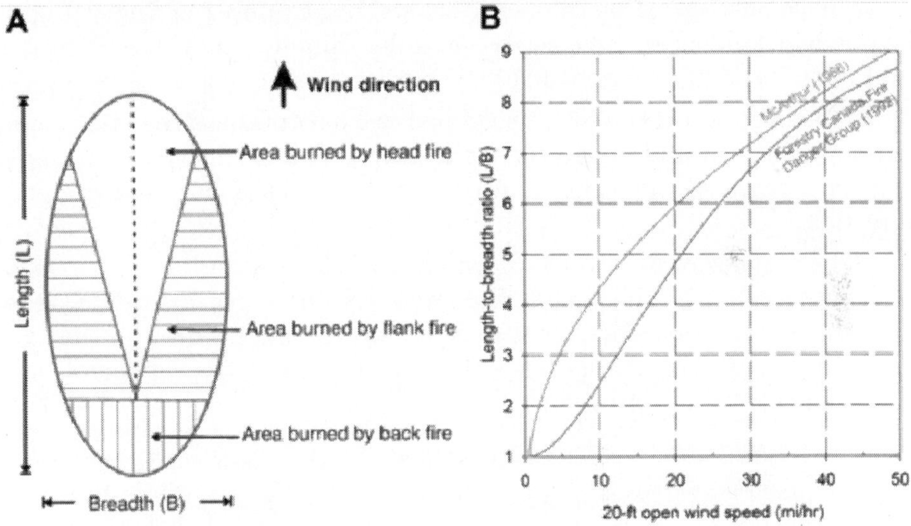

Figure : Fire elliptical shape under constant environment.

- A finger is a long, narrow extension of the main body of fire.
- A pocket is an unburned indentation of the fire perimeter surrounded on three sides by the fire.
- An island is an unburned area surrounded by burned area within fire.
- A spot fire is a fire initiated outside the main fire.

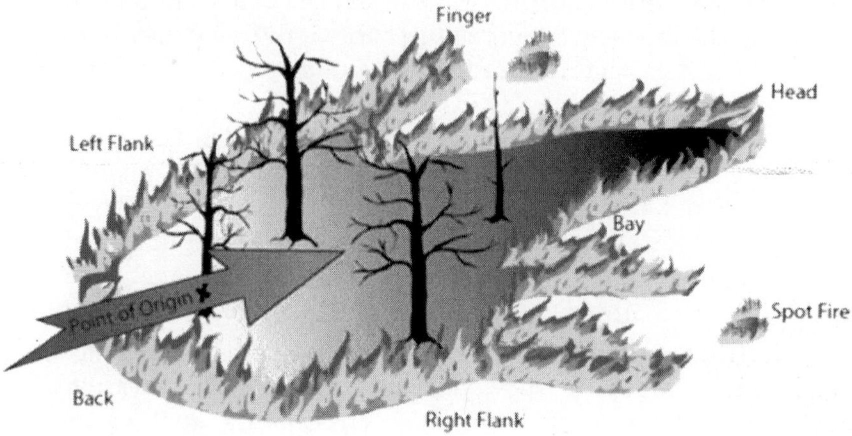

Figure : Parts of fire

These parts have no reference scales. For example, a finger might be one foot long or may exceed a one mile. Fingers might have fingers and spots might have spot fires.

The direction of the fire spread has two parameters: the maximum spread direction and long axis of the ellipse. The relative direction is the angle between the front flame orientation and the maximum spread direction.

Three types of fire propagation are as follows: ground fire, surface fire and crown fire. The perfect fire environment is a forest full of trees and shrubs, to start with surface fire then extend it through the tall trees to crown fire. At this stage, the fire will start to propagate, and it becomes hard to extinguish.

Figure : Crown fire

Monitoring systems

The information produced from monitoring can help the firefighters to understand the fire behaviour such as point of ignition, the spread speed and the direction of maximum spread. These parameters can be used as input for fire simulation programme to help extinguishing fire and provide safety to firefighter. Team work is very important for fire fighting to optimise the work with minimum efforts, minimum damage and the shortest time, for example, if the fire spread direction endangers human areas, the team priority target is to stop the propagation in that direction, so more firefighters should focus in that direction. On the contrary, to extinguish the fire with the shortest time, more firefighters can focus the team work on the mother/main fire not the finger to weaken the front flame, reduce the fire intensity, limit the spreading, reduce the dangerous and make it easier to extinguish.

Figure : Fire management

Fire suppression authorities and detection techniques

The common techniques used by authorities for fire detection and suppression :

1. controlled/supervised burning,
2. fire and weather casting and estimation for the fuel moisture level,
3. human watchtower,
4. optical smoke detection,
5. light detectors to detect and coordinate strikes,

6. infrared,
7. spotter planes,
8. water tankers,
9. mobile/smartphone calls.

Detection and monitoring systems are divided into two main groups:

1. volunteer reporting-public reporting of fires, public aircraft and ground-based field staff,
2. operational detection systems: fire towers, aerial patrols, electronic lightning detectors and automatic detection systems.

Many techniques used in fire suppression in different countries, like using flying water tankers like in Canada and burning dry region under the supervision of fire fighters rather than having a crisis later. Other countries sweep away the surrounding area to stop spreading and isolate it as in the Middle East. Australian authorities sometimes leave the fire to burn, until it dies alone, if it causes no harm for humans or properties.

Satellite system

Earth-orbiting satellites and even air-floating devices have been employed for the observation and detection of forest fires. Satellite images generated mainly by two satellites launched specifically for forest fire detection purposes—the advanced very high resolution radiometer (AVHRR) launched in 1998 and the moderate resolution imaging spectroradiometer (MODIS) launched in 1999—have been used. These satellites can provide images of the Earth once every 2 days and that is a long delay for fire scanning; in addition, the quality of satellite images depends on weather condition .

Any satellite-based observation for forest fires suffers from severe limitations resulting in a failure in the speed of detection or the quality or the running cost to produce effective control for forest areas. Some limitations in observation of forest fires from geostationary (GEO) or Low Earth Orbit (LEO) satellite are as follows: it is impossible to provide a full coverage by using satellite monitoring. GEO and LEO satellites are located on orbits over 22,800 miles above the Earth's surface. Infrared radiations emitted by fire flames may be too feeble in intensity to be detected by satellites at early stage. The intensity of radiation decreases with the square distance, and the radiation beam angle of arrival in regards to the surface of the detector or camera. However, satellites are not the best choice to detect the forest fires at early stages.

Some satellites might not be equipped with detection of forest fires, such as transponders, frequency translation, amplification reception, regeneration, antennas and downlink transmission. In fact, formal allocation for appropriate frequency and bandwidth for forest fire detection purposes is not available yet.

Optical sensor and digital camera

Camera surveillance and wireless sensor network are the other available techniques for forest fire detection. The technology advancement of cameras, image processing, industrial computers and sensors resulted in advance automated for early warning systems. Many types of detection are used in terrestrial systems :

1. video camera system is able to recognise a spectrum of smoke and fire during day and night.
2. thermal cameras detect the heat glow of fire.
3. IR spectrometers are used for the identification of spectral characteristics of smoke
4. LIDAR (detection of light and range) systems measure reflected laser rays from smoke particles.

All variety of optical camera systems used different algorithms designed by experts, where all rely on the same general concept (smoke

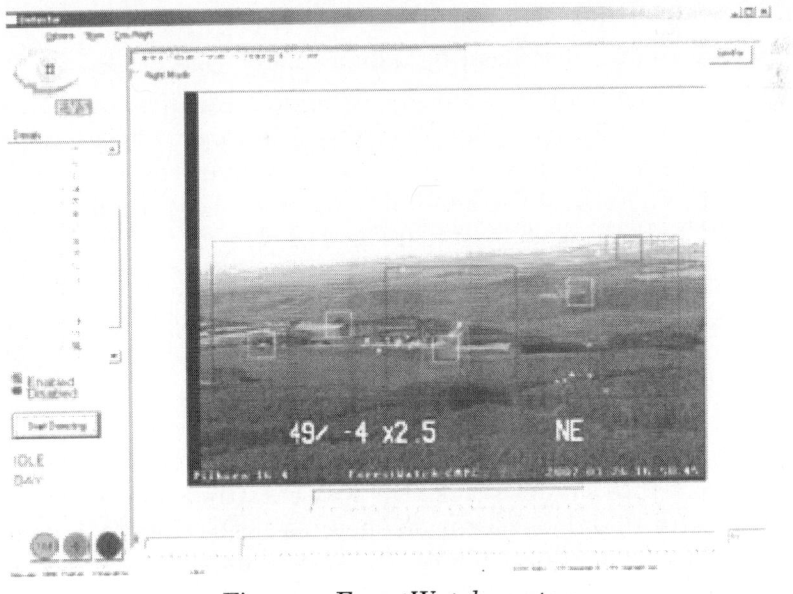

Figure : ForestWatch system

and fire glow detection). In general, cameras capture images every certain time, where images are number of pixels. The image-processing unit (1) tracks motion between sequenced images and (2) checks and compares pixels that contain smoke or fire glow. Then the results used in another algorithm to make a final decision whether the current situation requires action. Optical systems are integrated with geographical maps for localisation purposes.

Use of a given type of camera or sensor depends not only on the specific conditions of the operation but also on the financial resources available. The following are some available systems in the market:

AlarmEYE is a video and infrared camera system used for to detect early forest fire. It is based on black and white colour frequency where the infrared filter can distinguish between heat vapour and flames. This system was produced and deployed in Thailand .

EYEfi SPARC. Optical sensors are produced by EYEfi, Australia, for forest fire detection. Optical sensors consist of:

1. camera (ultralow light grey scale at night and colour during the day)
2. lightening detection sensor
3. communication unit
4. weather station
5. power system
6. tilt zoom cameras can be added to the system

EYEfi cannot provide a smoke detection, but it is on the future plans. Simply, EYEfi can play a helping role in providing images for fire agencies after the operator spots smoke by using the EYEfi software and GIS map to localise the smoke position on the ground. A weather station and lightening detector are included in the system for more accuracy.

UraFire system focused on reducing the number of false alarms by identifying smoke based on two inputs: (1) clustering motions and (2) a time input. The UraFire system is used and produced in France.

Forest Fire Finder is a unique, smart and complicated system based on atmosphere analysis instead of detecting fire glow or column of smoke. Forest Fire Finder tracks the way the atmosphere absorbs the sunlight. Naturally, the absorption depends on the chemical molecules or composition in the atmosphere. Different composition has different absorption behaviours; therefore, Forest Fire Finder can define different

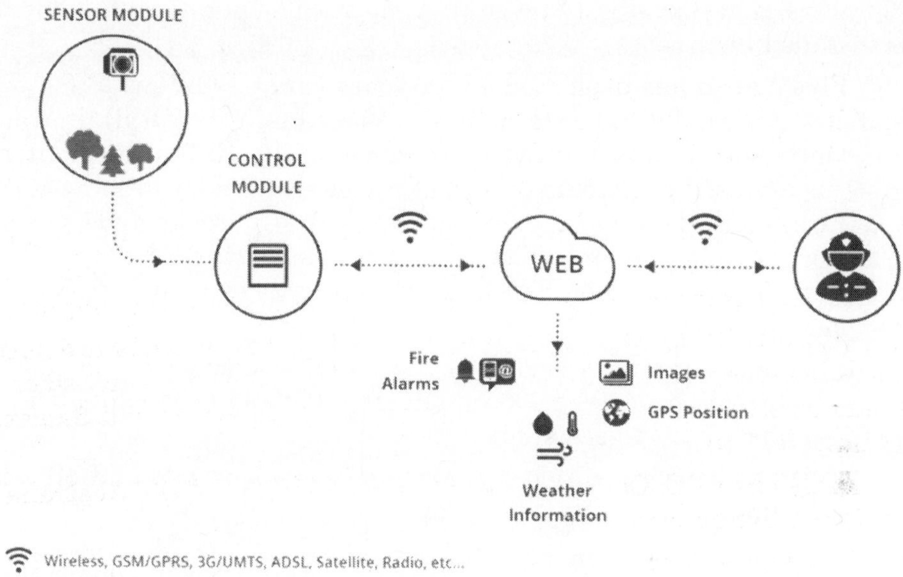

Figure : Forest fire finder

kinds of smoke based on the source in a range of 15 km, such as the organic smoke from burnt trees and the industrial smoke. The equipment installed in tree crowns for faster detection, and this system is used in Portuguese forests.

Forest Watch is an optical system based on sensor cameras, which provide a semiautomatic fire detection produced by Enviro-Vision Solutions, South Africa. The camera tower scans the forest in a range of 16–20 km for a column of smoke during day and fire glow during night. ForestWatch consists of:

1. a 360° Pan tilt cameras to allow +33 to “83 tilt from horizon, with 24× optical zoom,
2. telecommunications system such as microwave, 3G or satellite,
3. image sampling engine,
4. processing software to evaluate the current situation.

The most popular system in forest fire detection is ForestWatch. On the contrary, only Canada has documented its own experience with the system. Canada pointed that fires were detected in a range of 20 km reliably, but a large number of faulty alarms were also generated. Operational ForestWatch systems are in use in the USA (22 towers),

Canada (4 towers), South Africa (83 towers), Chile (20 towers), Swaziland (5 towers) and Slovakia (4 towers). A pilot scale operation (2 towers) is installed in Greece.

FireWatch has been studied for years (since 1992) in Germany, and now it is produced by German Aerospace Institute (DLR). It is an automatic smoke detection within a range of 10–40 km. FireWatch systems are used in Estonia (5 towers, 1 control room), Germany (178 towers, 22 control rooms), Mexico (1 tower, 1 control room) and Cyprus (2 towers, 1 control room). Pilot systems (1 or 2 towers) are in use in Italy, Czech Republic, the USA, Spain, Greece and Portugal . FireWatch components consist of :

1. Optical sensor system (OSS): the optical system performs a 360° rotation once every 4–6 min during day and 8–12 min during night in 10-degree steps.
2. Data transfer: wireless connection between OSS and officer computer.
3. Central office: the work space/area (computers, monitors and printer).

Figure : FireWatch

FireHawk. A risk management system that can localise fire consists of the following layers :

1. imaging layer: represents cameras installation on suitable positions,
2. communication layers: wireless links,
3. machine vision layer: FireHawk uses the GIS and the ForestWatch software to localise and find the shortest path to the fire. FireHawk is deployed in two areas in South Africa.

The Bushfire Cooperative Research Centre in collaboration with the Australian government published a very interesting report about the experience of using three optical systems. The target is to testify the performance of each one of the systems compared to normal human observation towers . The report provides full details about the project, environment, experiment, analysis and results. The three systems used in this project are FireWatch, EYEfi and ForestWatch in 2010, all of which had been tested on three kinds of fires: agency burning, private burning and research burning in Otway Ranges in Victoria and Tumut in NSW.

Figure : Tower in Tumut, Australia, with the three systems

1. More than 250 private fires were reported in the study area. A high proportion of them were reported by observation towers not by cameras.
2. Thirty-seven prescribed fires and burn-off fires were conducted as part of land management tasks during the post fire season in NSW forests, National Parks and Wildlife Service and Victorian

Department of Sustainability and Environment. One unplanned bushfire was reported in Victoria and none in NSW.

As a result of this project, the trained human tower observer is more reliable and faster than the three systems together, and it is not possible to rely on optical cameras only. However, it might be useful to improve the performance of human observer or cover unstaffed tower or remote areas. They indicate that FireWatch detected more fires than the other systems and the only system to report the forest burns, although the detection was 35 min later than the human tower detection and the fire was five times bigger. The system performance was partially a function of distance from camera and the size of the fire; moderate burns (10–20 km) were missed while large burns were detected at long distances (70 km). The systems are not able to take landscape topography in calculation, which leads to misallocation of the fire.

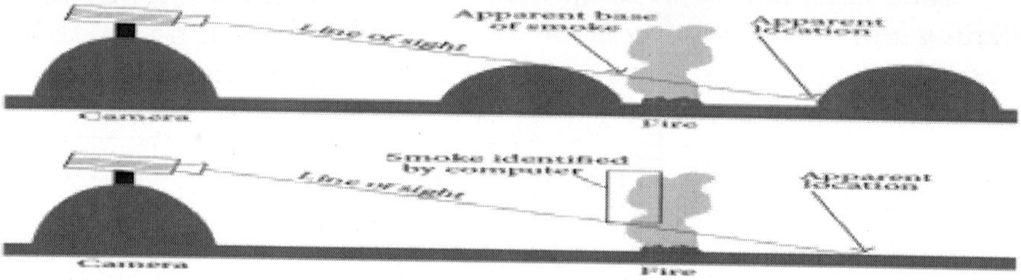

Figure : Localisation error reason

They concluded the report with the following points.

1. Cameras are not suitable replacements for trained human observation tower.
2. Human towers and optical systems were used for forest fire scanning for column of smoke. There is a delay between ignition time and a noticeable column of smoke that can be detected by optical cameras or human.
3. Landscape appearance, atmosphere conditions, and the properties of the smoke are different from one country to another.

Optical systems require more improved and intelligent algorithms and techniques in regard to the number of faulty alarms caused by various dynamic phenomena such as reflections, wind-tossed trees, human activity and cloud shadows. Processing landscape images is very difficult due to nature variety and to the dynamic events that may cause various illuminating, which depends on distance, weather, masking objects, time of day and so forth. These events produce dynamic envelopes, which are not always caused by motion and consist of time-varying grey levels of connected pixels in several image regions.

Optical cameras are designed to cover large areas (in a range of 15–80 km) with a minimum number of camera towers, and they can only provide a line of sight vision, which is a problem in the forest areas, as hills, high trees and mountains might block the vision. Weather condition and night vision reflect on the camera performance.

Camera surveillance technology with short distance links was also tried, despite the need for manual installation for every camera in a suitable position. Still the line of sight images, bad weather image, night images and high probability of false alarms for the following reasons: (1) daily motion of the Sun, (2) vegetation, (3) variation of atmospheric extinction and (4) moving clouds.

Finally, optical systems are extremely expensive; optical tower might worth more than $30,000 per tower, and there is a need to build these towers and install a communication infrastructure in the remote areas inside the forests.

Wireless sensor network

The line of sight problem of optical cameras in forests can be solved with the second type of sensors. A new technology called wireless sensor network (WSN) can be deployed in large number of systems, one potential application is forest fire detection. The same printed circuit board integrates the wireless transceiver, sensors and data processing. Sensors are able to influence the physical parameters such as pressure, temperature, gases, radiations, humidity and many other parameters. Sensor network normally deployed in a large-scale random distribution on remote or inaccessible places and under harsh environment for a certain period of lifetime. This technology relied on low data rate and short ranges of communication with multi-hope fashion to reach the sink.

The recent advancement in sensor network technology has made it possible to use this technology in early forest fire detection. A number of studies have considered using WSN in wood fire systems.

Spain used a sensor network and IP cameras to detect the fire ignition by sensors and use the closest camera to provide images for fire. Spain tried four IP cameras where they installed manually in the forest and images are heavy load on such a limited resources network such as sensor network.

Forest Fire Surveillance System (FFSS) is a South Korean system. The surveillance is done by sensor network to observe illumination, temperature and humidity. These readings go into a database to make a daily calculation and comparison in order to evaluate the hazards.

FireWxNet is a system target to study the forest fire behaviour not detection. The system uses wireless sensor network to provide data for weather status and web cameras to provide images for the fire. The system uses a tiered structure, which starts with directional antennas on the top of mountains and ends with multi-hope sensor network to observe the required environmental parameters. They used web cameras to provide vision data as well, and they equipped the sensors with a small GPS device to provide the location information.

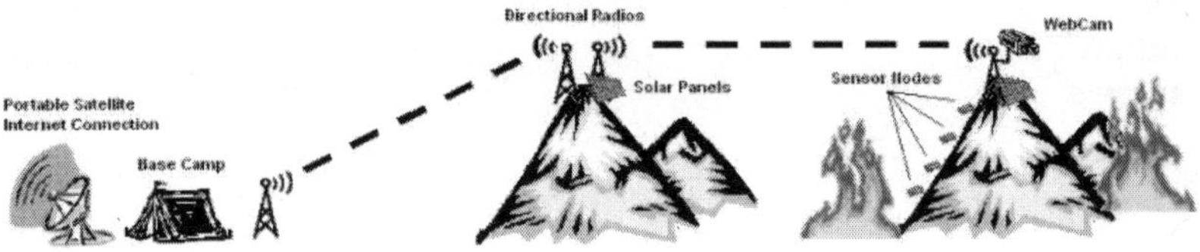

Figure : FireWxNet

It is a very smart system proposed in Canada. The system based on fire weather index (FWI) to calculate the probability of fire and the spread speed of the fire. The model provides the fire probability, spread speed, weather observation, moisture content and the fuel codes, which is divided into the following three types to describe the soil content of forest ground.

1. Fine Fuel Code (FFMC) represents the litter and fine fuels for 2 cm deep.
2. Duff Moisture Code (DMC) represents moisture content of decomposing organic material for 5–10 cm deep.
3. Drought Code (DC) represents the moisture organic content for 10–20 cm deep.

It is a Forest Fire Detection project in Pennsylvania. The system uses fire sensors and GPS devices. The project has two aims: (1) rely on the existing technology and (2) replace all the existing fire detection techniques with more efficient ones.

The project plan is to install 12,000 units within 48 months, 4000 devices every 12–15 months. When the sensor detects fire or smoke, a signal is sent through GPS device to Satellite, and then Satellite will forward the signal to monitoring screen and other handheld devices.

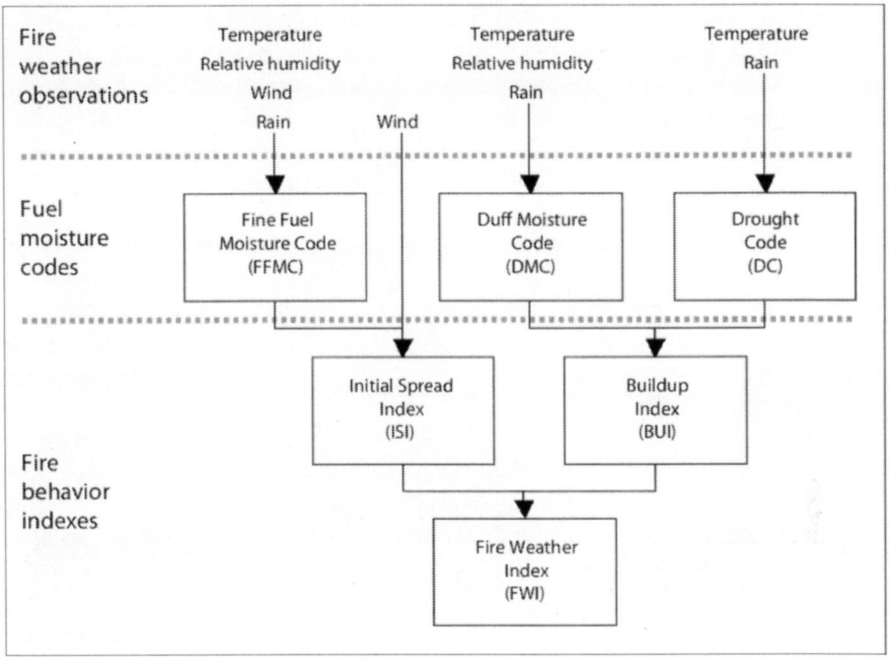

Figure : FWI system

Figure : Pennsylvania project

Inner magnolia forest fire research has been done by a system of three parts: (1) monitoring, (2) information management and database system and (3) decision-making system. The system provides a fire simulation from the field images in 3D maps by using Geodatabase and ArcSDE programs for fire simulation.

FIRESENSE (Fire Detection and Management through a Multisensor Network for the Protection of Cultural Heritage Areas from the Risk of Fire and Extreme Weather Conditions, FP7-ENV-2009-1-244,088-FIRESENSE) is a Specific Targeted Research Project of the European Union's 7th Framework Program Environment (including climate change). The FIRESENSE FP7 is a target to monitor remote areas and provide warning system. FIRESENSE is a very advanced system; it relies on IR, optical, temperature sensors, PTZ cameras and weather stations. All these sensors collect and process data to provide a clear understanding for the event to the local authority. The project deployments will be in Turkey, Italy Tunisia and Greece.

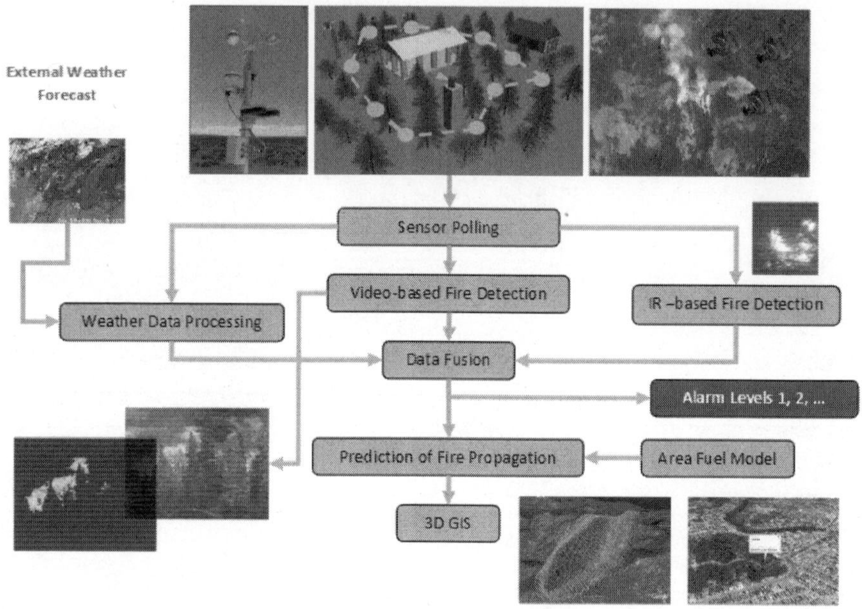

Figure : FIRESENSE FP7.

FP7 relies on complicated models, algorithms, concepts and comparisons. They are given as follows:

1. Scene model (Planck's radiation formula): the heat flux, thermal emitting, smoke, the fire, the reflectance, flickering, absorption emission lines, analysis of the atoms (e.g. potassium) and the molecules (water and carbon dioxide) are characteristics to be investigated.
2. Thermal heat emitted from the background, sunlight reflection, clouds shadow, the buildings and the sky polarisation.
3. Atmosphere gases (N2, O_2, CO, CO_2, H_2O, etc.); each gas behaves and absorbs differently; water vapour concentration; carbon dioxide is more uniformly distributed—its value is larger over industrial cities and vegetation fields than over oceans and deserts.

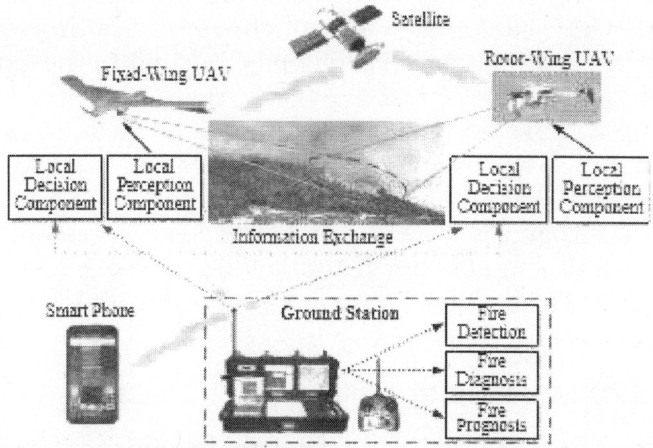

Figure : How FIRESENSE analyse images.

Libelium is a Spanish wireless sensor network company. They named their product Waspmote and proposed it for many WSN applications such as forest fire detection, smart cities, water pollution and many other applications. With regard to forest fire detection, they used Waspmote nodes equipped with GPS device for localisation, and gas boards to measure temperature, carbon dioxide (CO_2), carbon monoxide (CO) and humidity for detection. Libelium deployed 90 nodes with solar panels for power scavenging to measure parameters every 5 min.

Figure : Waspmote and gas board

Wireless sensor network technology usually deployed in large number that can observe and the surrounding environment, transforming it into electrical signals, to send to the sink in a multi-hop fashion for processing. By this way, there is no need to build towers or set up complicated communication links such as microwave and satellite. WSN works on short communication links fashion and can provide real-time monitoring, where using this technology for forest fire application requires a large number of randomly deployed nodes to provide a reliable network if the key issues were addressed for this network: (1) localisation, (2) coverage, (3) network life span and (4) fire detection method.

Forest Fire Monitoring

Forest fires cause billions of dollars in damage to property and the environment every year. To combat forest fires effectively, their early detection and continuous tracking is vital.

With the help of advanced image processing techniques, many methods have been developed to detect a forest fire in remote regions using satellite images. Such images are typically taken by low earth orbiting satellites with an orbital period of about ten hours, and with a resolution that is only sufficient for fire detection. However, fire fighters need frequent and high-quality information updates of a fire's development in order to conduct an effective and safe fire fighting mission. Because forest fire monitoring is a difficult task, fire fighters often have to enter a fire region with little knowledge of how and where the fire is propagating, placing their lives at risk. For these reasons, there is an urgent need to develop more effective fire monitoring technologies.

Low-altitude, short-endurance (LASE) UAVs are expected to be a key technology for closing the fire monitoring gap. Flying at low altitude, these UAVs can capture high resolution imagery and broadcast frequent updates to fire crews. NASA is actively pursuing this possibility with ongoing research projects aimed at tracking the growth of fires using LASE UAVs . However, a number of challenges have to be solved before LASE UAVs can be used for fire monitoring. First, with the fire growing and changing directions, UAVs need to be able to plan their path using limited real-time information. Second, LASE UAVs typically cannot carry enough fuel to endure a long fire fighting mission, which means the UAV needs to have the intelligence to know when to return to base for refueling.

Furthermore, for large forest fires, the information update rate may still be too low if only a single LASE UAV is employed.

The world we live in has experienced several ecological disasters, most of them caused by direct human actions, which have been intensifying particularly in the past few decades. In particular, forest fire is a natural hazard, whose intensity and frequency have recently been increased, thus affecting forest ecosystems with serious consequences on the environment, society and economy. In addition, drier, warmer weather caused by climate change creates conditions for extreme and widespread biomass burning. Spaceborne and airborne remote sensing Earth observation platforms with improved instruments (including optical, thermal infrared, microwave sensors) have rapidly evolved in the last decades, thus providing an invaluable data source, to understand, monitor and respond to wildfires, at various spatial and temporal scales. Optical and SAR data with global coverage and frequent revisits have become freely available, thus playing a critical role in wildfire monitoring. Moreover, LiDAR has the potential to penetrate the forest canopy, thus emerging as a powerful active sensing tool for assessing both landscape fire hazard and fire effects on vegetation at fine-grained scale. Together to remote sensing analyses, experimental analyses of fire effects are important to have a complementary information about the impact of fire.

CHAPTER 7

PLANNING, ORGANIZATION AND TECHNIQUES OF FOREST FIRE CONTROL

Detection

A key aspect of fire protection is to identify a developing fire emergency in a timely manner, and to alert the building's occupants and fire emergency organizations. This is the role of fire detection and alarm systems. Depending on the anticipated fire scenario, building and use type, number and type of occupants, and criticality of contents and mission, these systems can provide several main functions. First they provide a means to identify a developing fire through either manual or automatic methods and second, they alert building occupants to a fire condition and the need to evacuate. Another common function is the transmission of an alarm notification signal to the fire department or other emergency response organization. They may also shut down electrical, air handling equipment or special process operations, and they may be used to initiate automatic suppression systems. This section will describe the basic aspects of fire detection and alarm systems.

Fire Growth and Behavior

Before attempting to understand fire detection systems and automatic sprinklers, it is beneficial to possess a basic knowledge of fire development and behavior. With this information, the role and interaction of these supplemental fire safety systems in the protection process can then be better realized.

Basically, a fire is a chemical reaction in which a carbon based material (fuel), mixes with oxygen (usually as a component of air), and is heated to a point where flammable vapors are produced. These vapors can then come in contact with something that is hot enough to cause vapor ignition, and a resulting fire. In simple terms, something that can burn touches something that is hot, and a fire is produced.

Libraries, archives, museums, and historic structures frequently contain numerous fuels. These include books, manuscripts, records, artifacts, combustible interior finishes, cabinets, furnishings, and laboratory chemicals. It should be recognized that any item containing wood, plastic, paper, fabric, or combustible liquids is a potential fuel. They also contain several common, potential ignition sources including any item, action, or process which produces heat. These encompass electric lighting and power systems, heating and air conditioning equipment, heat producing conservation and maintenance activities, and electric office appliance. Flame generating construction activities such as soldering, brazing, and cutting are frequent sources of ignition. Arson is unfortunately one of the most common cultural property ignition sources, and must always be considered in fire safety planning.

When the ignition source contacts the fuel, a fire can start. Following this contact, the typical accidental fire begins as a slow growth, smoldering process which may last from a few minutes to several hours. The duration of this "incipient" period is dependent on a variety of factors including fuel type, its physical arrangement, and quantity of available oxygen. During this period heat generation increases, producing light to moderate volumes of smoke. The characteristic smell of smoke is usually the first indication that an incipient fire is underway. It is during this stage that early detection (either human or automatic), followed by a timely response by qualified fire emergency professionals, can control the fire before significant losses occur.

As the fire reaches the end of the incipient period, there is usually enough heat generation to permit the onset of open, visible flames. Once flames have appeared, the fire changes from a relatively minor situation to a serious event with rapid flame and heat growth. Ceiling temperatures can exceed 1,000° C (1,800° F) within the first minutes. These flames can ignite adjacent combustible contents within the room, and immediately endanger the lives of the room's occupants. Within 3–5 minutes, the room ceiling acts like a broiler, raising temperatures high enough to "flash", which simultaneously ignites all combustibles in the room. At this point, most contents will be destroyed and human survivability becomes impossible. Smoke generation in excess of several thousand cubic meters (feet) per minute will occur, obscuring visibility and impacting contents remote from the fire.

If the building is structurally sound, heat and flames will likely consume all remaining combustibles and then self extinguish (burn out). However, if wall and/or ceiling fire resistance is inadequate, (i.e. open doors, wall/ceiling breaches, combustible building construction), the fire can spread into adjacent spaces, and start the process over. If the fire remains uncontrolled, complete destruction or "burn out" of the entire building and contents may ultimately result.

Successful fire suppression is dependent on extinguishing flames before, or immediately upon, flaming combustion. Otherwise, the resulting damage may be too severe to recover from. During the incipient period, a trained person with portable fire extinguishers may be an effective first line of defense. However, should an immediate response fail or the fire grow rapidly, extinguisher capabilities can be surpassed within the first minute. More powerful suppression methods, either fire department hoses or automatic systems, then become essential.

A fire can have far reaching impact on the institution's buildings, contents and mission. General consequences may include:

- Collections damage. Most heritage institutions house unique and irreplaceable objects. Fire generated heat and smoke can severely damage or totally destroy these items beyond repair.
- Operations and mission damage. Heritage occupancies often contain educational facilities, conservation laboratories, catalogue services, administrative/support staff offices, exhibition production, retail, food service, and a host of other activities. A fire can shut these down with adverse impact on the organization's mission and its clientele.
- Structure damage. Buildings provide the "shell" that safeguards collections, operations and occupants from weather, pollution, vandalism and numerous other environmental elements. A fire can destroy walls, floors, ceiling/roof assemblies and structural support, as well as systems that illuminate, control temperature and humidity, and supply electrical power. This can in turn lead to content harm, and expensive relocation activities.
- Knowledge loss. Books, manuscripts, photographs, films, recordings and other archival collections contain a vast wealth of information that can be destroyed by fire.
- Injury or loss of life. The lives of staff and visitors can be endangered.
- Public relations impact. Staff and visitors expect safe conditions

in heritage buildings. Those who donate or loan collections presume these items will be safeguarded. A severe fire could shake public confidence and cause a public relations impact.

- Building security. A fire represents the single greatest security threat! Given the same amount of time, an accidental or intentionally set fire can cause far greater harm to collections than the most accomplished thieves. Immense volumes of smoke and toxic gases can cause confusion and panic, thereby creating the ideal opportunity for unlawful entry and theft. Unrestricted firefighting operations will be necessary, adding to the security risk. Arson fires set to conceal a crime are common.

To minimize fire risk and its impact, heritage institutions should develop and implement comprehensive and objective fire protection programs. Program elements should include fire prevention efforts, building construction improvements, methods to detect a developing fire and alert emergency personnel, and means to effectively extinguish a fire. Each component is important toward overall accomplishment of the institution's fire safety goal. It is important for management to outline desired protection objectives during a fire and establish a program that addresses these goals. Therefore, the basic question to be asked by the property's managers is, "What maximum fire size and loss can the institution accept?" With this information, goal oriented protection can be implemented.

Fire detectors sense one or more of the products or phenomena resulting from fire, such as smoke, heat, infrared and/or ultraviolet light radiation, or gas.

In dwellings, smoke detectors are often stand-alone devices. In non-domestic buildings, fire detection will typically take the form of a fire alarm system, incorporating one or more of the following automatic devices:

- Heat detector
- Smoke detector
- Flame detector
- Fire gas detector

Control Panels

The control panel is the "brain" of the fire detection and alarm system. It is responsible for monitoring the various alarm "input" devices such as manual and automatic detection components, and then activating alarm "output" devices such as horns, bells, warning lights, emergency telephone dialers, and building controls. Control panels may range from simple units with a single input and output zone, to complex computer driven systems that monitor several buildings over an entire campus. There are two main control panel arrangements, conventional and addressable, which will be discussed below.

Conventional or "point wired" fire detection and alarm systems were for many years the standard method for providing emergency signaling. In a conventional system one or more circuits are routed through the protected space or building. Along each circuit, one or more detection devices are placed. Selection and placement of these detectors is dependent upon a variety of factors including the need for automatic or manual initiation, ambient temperature and environmental conditions, the anticipated type of fire, and the desired speed of response. One or more device types are commonly located along a circuit to address a variety of needs and concerns.

Upon fire occurrence, one or more detectors will operate. This action closes the circuit, which the fire control panel recognizes as an emergency condition. The panel will then activate one or more signaling circuits to sound building alarms and summon emergency help. The panel may also send the signal to another alarm panel so that it can be monitored from a remote point.

In order to help insure that the system is functioning properly, these systems monitor the condition of each circuit by sending a small current through the wires. Should a fault occur, such as due to a wiring break, this current cannot proceed and is registered as a "trouble" condition. The indication is a need for service somewhere along the respective circuit.

In a conventional alarm system, all alarm initiating and signaling is accomplished by the system's hardware which includes multiple sets of wire, various closing and opening relays, and assorted diodes. Because of this arrangement, these systems are actually monitoring and controlling circuits, and not individual devices.

To further explain this, assume that a building's fire alarm system has 5 circuits, zones A through E, and that each circuit has 10 smoke detectors and 2 manual stations located in various rooms of each zone. A fire ignition in one of the rooms monitored by zone "A" causes a smoke detector to go into alarm. This will be reported by the fire alarm control panel as a fire in circuit or zone "A". It will not indicate the specific detector type nor location within this zone. Emergency responding personnel may need to search the entire zone to determine where the device is reporting a fire. Where zones have several rooms, or concealed spaces, this response can be time consuming and wasteful of valuable response opportunity.

The advantage of conventional systems is that they are relatively simple for small to intermediate size buildings. Servicing does not require a large amount of specialized training.

A disadvantage is that for large buildings, they can be expensive to install because of the extensive amounts of wire that are necessary to accurately monitor initiating devices.

Conventional systems may also be inherently labor intensive and expensive to maintain. Each detection device may require some form of operational test to verify it is in working condition. Smoke detectors must be periodically removed, cleaned, and recalibrated to prevent improper operation. With a conventional system, there is no accurate way of determining which detectors are in need of servicing. Consequently, each detector must be removed and serviced, which can be a time consuming, labor intensive, and costly endeavor. If a fault occurs, the "trouble" indication only states that the circuit has failed, but does not specifically state where the problem is occurring. Subsequently, technicians must survey the entire circuit to identify the problem.

Addressable or "intelligent" systems represent the current state-of-the-art in fire detection and alarm technology. Unlike conventional alarm methods, these systems monitor and control the capabilities of each alarm initiating and signaling device through microprocessors and system software. In effect, each intelligent fire alarm system is a small computer overseeing and operating a series of input and output devices.

Like a conventional system, the address system consists of one or more circuits that radiate throughout the space or building. Also, like standard systems, one or more alarm initiating devices may be located along these circuits. The major difference between system types involves the way in which each device is monitored. In an addressable system, each initiating device (automatic detector, manual station, sprinkler waterflow switch, etc.) is given a specific identification or "address". This address is correspondingly programmed into the control panel's memory with information such as the type of device, its location, and specific response details such as which alarm devices are to be activated.

The control panel's microprocessor sends a constant interrogation signal over each circuit, in which each initiating device is contacted to inquire its status (normal or emergency). This active monitoring process occurs in rapid succession, providing system updates every 5 to 10 seconds.

The addressable system also monitors the condition of each circuit, identifying any faults which may occur. One of the advancements offered by these systems is their ability to specifically identify where a fault has developed. Therefore, instead of merely showing a fault along a wire, they will indicate the location of the problem. This permits faster diagnosis of the trouble, and allows a quicker repair and return to normal.

Advantages provided by addressable alarm systems include stability, enhanced maintenance, and ease of modification. Stability is achieved by the system software. If a detector recognizes a condition which could be indicative of a fire, the control panel will first attempt a quick reset. For most spurious situations such as insects, dust, or breezes, the incident will often remedy itself during this reset procedure, thereby reducing the probability of false alarm. If a genuine smoke or fire condition exists, the detector will reenter the alarm mode immediately after the reset attempt. The control panel will now regard this as a fire condition, and will enter its alarm mode.

With respect to maintenance, these systems offer several key advantages over conventional ones. First of all, they are able to monitor the status of each detector. As a detector becomes dirty, the microprocessor recognizes a decreased capability, and provides a maintenance alert. This feature, known as Listed Integral Sensitivity Testing, allows facilities personnel to service only those detectors that needs attention, rather than requiring a labor and time consuming cleaning of all units.

Advanced systems, such as the FCI 7200 incorporate another maintenance feature known as drift compensation. This software procedure adjusts the detector's sensitivity to compensate for minor dust conditions. This avoids the ultra sensitive or "hot" detector condition which often results as debris obscures the detector's optics. When the detector has been compensated to its limit, the control panel alerts maintenance personnel so that servicing can be performed.

Modifying these systems, such as to add or delete a detector, involves connecting or removing the respective device from the addressable circuit, and changing the appropriate memory section. This memory change is accomplished either at the panel or on a personal computer, with the information downloaded into the panel's microprocessor.

The main disadvantage of addressable systems is that each system has its own unique operating characteristics. Therefore, service technicians must be trained for the respective system. The training program is usually a 3-4 day course at the respective manufacturer's facility. Periodic update training may be necessary as new service methods are developed.

Fire Detectors

When present, humans can be excellent fire detectors. The healthy person is able to sense multiple aspects of a fire including the heat, flames, smoke, and odors. For this reason, most fire alarm systems are designed with one or more manual alarm activation devices to be used by the person who discovers a fire. Unfortunately, a person can also be an unreliable detection method since they may not be present when a fire starts, may not raise an alarm in an effective manner, or may not be in perfect heath to recognize fire signatures. It is for this reason that a variety of automatic fire detectors have been developed. Automatic detectors are meant to imitate one or more of the human senses of touch, smell or sight. Thermal detectors are similar to our ability to identify high temperatures, smoke detectors replicate the sense of smell, and flame detectors are electronic eyes. The properly selected and installed automatic detector can be a highly reliable fire sensor.

Manual fire detection is the oldest method of detection. In the simplest form, a person yelling can provide fire warning. In buildings, however, a person's voice may not always transmit throughout the structure. For this reason, manual alarm stations are installed. The general design philosophy is to place stations within reach along paths of escape. It is for this reason that they can usually be found near exit doors in corridors and large rooms.

The advantage of manual alarm stations is that, upon discovering the fire, they provide occupants with a readily identifiable means to activate the building fire alarm system. The alarm system can then serve in lieu of the shouting person's voice. They are simple devices, and can be highly reliable when the building is occupied. The key disadvantage of manual stations is that they will not work when the building is unoccupied. They may also be used for malicious alarm activations. Nonetheless, they are an important component in any fire alarm system.

Thermal detectors are the oldest type of automatic detection device, having origin in the mid 1800's, with several styles still in production today. The most common units are fixed temperature devices that operate when the room reaches a predetermined temperature (usually in the 135°–165°F/57°–74°C). The second most common type of thermal sensor is the rate-of-rise detector, which identifies an abnormally fast temperature climb over a short time period. Both of these units are "spot type" detectors, which means that they are periodically spaced along a ceiling or high on a wall. The third detector type is the fixed temperature line type detector, which consists of two cables and an insulated sheathing that is designed to breakdown when exposed to heat. The advantage of line type over spot detection is that thermal sensing density can be increased at lower cost.

Thermal detectors are highly reliable and have good resistance to operation from nonhostile sources. They are also very easy and inexpensive to maintain. On the down side, they do not function until room temperatures have reached a substantial temperature, at which point the fire is well underway and damage is growing exponentially. Subsequently, thermal detectors are usually not permitted in life safety applications. They are also not recommended in locations where there is a desire to identify a fire before substantial flames occur, such as spaces where high value thermal sensitive contents are housed.

Smoke detectors are a much newer technology, having gained wide usage during the 1970's and 1980's in residential and life safety applications. As the name implies, these devices are designed to identify a fire while in its smoldering or early flame stages, replicating the human sense of smell. The most common smoke detectors are spot type units, that are placed along ceilings or high on walls in a manner similar to spot thermal units. They operate on either an ionization or photoelectric principle, with each type having advantages in different applications. For large open spaces such as galleries and atria, a frequently used smoke detector is a projected beam unit. This detector consists of two components, a light transmitter and a receiver, that are mounted at some distance (up to 300 ft/100m) apart. As smoke migrates between the two components, the transmitted light beam becomes obstructed and the receiver is no longer able to see the full beam intensity. This is interpreted as a smoke condition, and the alarm activation signal is transmitted to the fire alarm panel.

A third type of smoke detector, which has become widely used in extremely sensitive applications, is the air aspirating system. This device consists of two main components: a cotrol unit that houses the detection chamber, an aspiration fan and operation circuitry; and a network of sampling tubes or pipes. Along the pipes are a series of ports that are designed to permit air to enter the tubes and be transported to the detector. Under normal conditions, the detector constantly draws an air sample into the detection chamber, via the pipe network. The sample is analyzed for the existence of smoke, and then returned to atmosphere. If smoke becomes present in the sample, it is detected and an alarm signal is transmitted to the main fire alarm control panel. Air aspirating detectors are extremely sensitive and are typically the fastest responding automatic detection method. Many high technology organizations, such as telephone companies, have standardized on aspiration systems. In cultural properties they are used for areas such as collections storage vaults and highly valuable rooms. These are also frequently used in aesthetically sensitive applications since components are often easier to conceal, when compared to other detection methods.

The key advantage of smoke detectors is their ability to identify a fire while it is still in its incipient. As such, they provide added opportunity for emergency personnel to respond and control the developing fire before severe damage occurs. They are usually the preferred detection method in life safety and high content value applications. The disadvantage of smoke detectors is that they are usually more expensive to install, when compared to thermal sensors, and are more resistant to inadvertent alarms. However, when properly selected and designed, they can be highly reliable with a very low probability of false alarm.

Flame detectors represent the third major type of automatic detection method, and imitate the human sense of sight. They are line of sight devices that operate on either an infrared, ultraviolet or combination principle. As radiant energy in the approximate 4,000 to 7,700 angstroms range occurs, as indicative of a flaming condition, their sensing equipment recognizes the fire signature and sends a signal to the fire alarm panel.

The advantage of flame detection is that it is extremely reliable in a hostile environment. They are usually used in high value energy and transportation applications where other detectors would be subject to spurious activation. Common uses include locomotive and aircraft maintenance facilities, refineries and fuel loading platforms, and mines. A disadvantage is that they can be very expensive and labor intensive to maintain. Flame detectors must be looking directly at the fire source, unlike thermal and smoke detectors which can identify migrating fire signatures. Their use in cultural properties is extremely limited.

Alarm Output Devices

Upon receiving an alarm notification, the fire alarm control panel must now tell someone that an emergency is underway. This is the primary function of the alarm output aspect of a system. Occupant signaling components include various audible and visual alerting components, and are the primary alarm output devices. Bells are the most common and familiar alarm sounding device, and are appropriate for most building applications. Horns are another option, and are especially well suited to areas where a loud signal is needed such as library stacks, and architecturally sensitive buildings where devices need partial concealment. Chimes may be used where a soft alarm tone is preferred, such as health care facilities and theaters. Speakers are the fourth alarm sounding option, which sound a reproducible signal such as a recorded voice message. They are often ideally suited for large, multistory or other similar buildings where phased evacuation is preferred. Speakers also offer the added flexibility of emergency public address announcements. With respect to visual alert, there are a number of strobe and flashing light devices. Visual alerting is required in spaces where ambient noise levels are high enough to preclude hearing sounding equipment, and where hearing impaired occupants may be found. Standards such as the Americans with Disabilities Act (ADA) mandate visual devices in numerous museum, library, and historic building applications.

Another key function of the output function is emergency response notification. The most common arrangement is an automatic telephone or radio signal that is communicated to a constantly staffed monitoring center. Upon receiving the alert, the center will then contact the appropriate fire department, providing information about the location of alarm. In some instances, the monitoring station may be the police or fire departments, or a 911 center. In other instances it will be a private monitoring company that is under contract to the organization. In many cultural properties, the building's inhouse security service may serve as the monitoring center.

Other output functions include shutting down electrical equipment such as computers, shutting off air handling fans to prevent smoke migration, and shutting down operations such as chemical movement through piping in the alarmed area. They may also activate fans to extract smoke, which is a common function in large atria spaces. These systems can also activate discharge of gaseous fire extinguishing systems, or preaction sprinkler systems.

Summary

In summary, there are several options for a building's fire detection and alarm system. The ultimate system type, and selected components, will be dependent upon the building construction and value, its use or uses, the type of occupants, mandated standards, content value, and mission sensitivity. Contacting a fire engineer or other appropriate professional who understands fire problems and the different alarm and detection options is usually a preferred first step to find the best system.

Fire Sprinklers

For most fires, water represents the ideal extinguishing agent. Fire sprinklers utilize water by direct application onto flames and heat, which causes cooling of the combustion process and prevents ignition of adjacent combustibles. They are most effective during the fire's initial flame growth stage, while the fire is relatively easy to control. A properly selected sprinkler will detect the fire's heat, initiate alarm, and begin suppression within moments after flames appear. In most instances sprinklers will control fire advancement within a few minutes of their activation, which will in turn result in significantly less damage than otherwise would happen without sprinklers.

Among the potential benefits of sprinklers are the following:

- Immediate identification and control of a developing fire. Sprinkler systems respond at all times, including periods of low occupancy. Control is generally instantaneous.
- Immediate alert. In conjunction with the building fire alarm system, automatic sprinkler systems will notify occupants and emergency response personnel of the developing fire.
- Reduced heat and smoke damage. Significantly less heat and smoke will be generated when the fire is extinguished at an early stage.
- Enhanced life safety. Staff, visitors and fire fighters will be subject to less danger when fire growth is checked.
- Design flexibility. Egress route and fire/smoke barrier placement becomes less restrictive since early fire control minimizes demand on these systems. Many fire and building codes will permit design and operations flexibility based on the presence of a fire sprinkler system.

- Enhanced security. A sprinkler controlled fire can reduce demand on security forces by minimizing intrusion and theft opportunities.
- Decreased insurance expenditure. Sprinkler controlled fires are less damaging than fires in nonsprinklered buildings. Insurance underwriters may offer reduced premiums in sprinkler protected properties.

These benefits should be considered when deciding on the selection of automatic fire sprinkler protection.

Sprinkler System Components and Operation

Sprinkler systems are essentially a series of water pipes that are supplied by a reliable water supply. At selected intervals along these pipes are independent, heat activated valves known as sprinkler heads. It is the sprinkler that is responsible for water distribution onto the fire. Most sprinkler systems also include an alarm to alert occupants and emergency forces when sprinkler activation (fire) occurs.

During the incipient fire stage, the heat output is relatively low and is unable to cause sprinkler operation. However, as the fire intensity increases, the sprinkler's sensing elements become exposed to elevated temperatures (typically in excess of 57–107°C (135–225°F), and begin to deform. Assuming temperatures remain high, as they would during an increasing fire, the element will fatigue after an approximate 30 to 120 second period. This releases the sprinkler's seals allowing water to discharge onto the fire and begin the suppression action. In most situations less than 2 sprinklers are needed to control the fire. In fast growing fire scenarios, however, such as a flammable liquid spill, up to 12 sprinklers may be required.

In addition to normal fire control efforts, sprinkler operation may be interconnected to initiate building and fire department alarms, shutdown electrical and mechanical equipment, close fire doors and dampers, and suspend some processes.

As fire fighters arrive their efforts will focus on ensuring that the system has contained the fire, and, when satisfied, shut off the water flow to minimize water damage. It is at this point that staff will normally be permitted to enter the damaged space and perform salvage duties.

System Components and Types

The basic components of a sprinkler system are the sprinklers, system piping, and a dependable water source. Most systems also require an alarm, system control valves, and means to test the equipment.

The sprinkler itself is the spray nozzle, which distributes water over a defined fire hazard area (typically 14–21 m2/150–225 ft2) with each sprinkler operating by actuation of its own temperature linkage. The typical sprinkler consists of a frame, thermal operated linkage, cap, orifice, and deflector. Styles of each component may vary but the basic principles of each remain the same.

- **Frame.** The frame provides the main structural component which holds the sprinkler together. Water supply piping is connected to the sprinkler at the base of the frame. The frame holds the thermal linkage and cap in place, and supports the deflector during discharge. Frame styles include standard and low profile, flush, and concealed mount. Some are designed for extended spray coverage, beyond the range of normal sprinklers. Standard finishes include brass, chrome, black, and white, while custom finishes are available for aesthetically sensitive spaces. Special coatings are available for areas subject to high corrosive effect. Selection of a specific frame style is dependent on the size and type of area to be covered, anticipated hazard, visual impact features, and atmospheric conditions.
- **Thermal linkage.** The thermal linkage is the component that controls water release. Under normal conditions the linkage holds the cap in place and prevents water flow. As the link is exposed to heat, however, it weakens and releases the cap. Common linkage styles include soldered metal levers, frangible glass bulbs, and solder pellets. Each link style is equally dependable.

Upon reaching the desired operating temperature, an approximate 30 second to 4 minute time lag will follow. This lag is the time required for linkage fatigue and is largely controlled by the link materials and mass. Standard responding sprinklers operate closer to the 3–4 minute mark while quick response (QR) sprinklers operate in significantly shorter periods. Selection of a sprinkler response characteristic is dependent upon the existing risk, acceptable loss level, and desired response action.

In heritage applications the advantage of quick response sprinklers often becomes apparent. The faster a sprinkler reacts to a fire, the sooner the suppression activity is initiated, and the lower the potential damage level. This is particularly beneficial in high value or life safety applications where the earliest possible extinguishment is a fire protection goal. It is important to understand that response time is independent of response temperature. A quicker responding sprinkler will not activate at a lower temperature than a comparable standard head.

- **Cap.** The cap provides the water tight seal which is located over the sprinkler orifice. It is held in place by the thermal linkage, and falls from position after linkage heating to permit water flow. Caps are constructed solely of metal or a metal with a teflon disk.
- **Orifice.** The machined opening at the base of the sprinkler frame is the orifice from which extinguishing water flows. Most orifice openings are 15 mm (1/2 inch) diameter with smaller bores available for residential applications and larger openings for higher hazards.
- **Deflector.** The deflector is mounted on the frame opposite the orifice. Its purpose is to break up the water stream discharging from the orifice into a more efficient extinguishing pattern. Deflector styles determine how the sprinkler is mounted, with common sprinkler mounting styles known as upright (mounted above the pipe), pendent (mounted below the pipe, i.e. under ceilings), and sidewall sprinklers which discharge water in a lateral position from a wall. The sprinkler must be mounted as designed to ensure proper action. Selection of a particular style is often dependent upon physical building constraints.

A sprinkler that has received wide spread interest for museum applications is the on/off sprinkler. The principle behind these products is that as a fire occurs, water discharge and extinguishing action will happen similar to standard sprinklers. As the room temperature is cooled to a safer level, a bimetallic snap disk on the sprinkler closes and water flow ceases. Should the fire reignite, operation will once again occur. The advantage of on/off sprinklers is their ability to shut off, which theoretically can reduce the quantity of water distributed and resultant damage levels. The problem, however, is the long time period that may pass before room temperatures are sufficiently cooled to the sprinkler's shut off point. In most heritage applications, the building's construction will retain heat and prevent the desired sprinkler shut down. Frequently, fire emergency response forces will have arrived and will be able to close sprinkler zone control valves before the automatic shut down feature has functioned.

On-off sprinklers typically cost 8–10 times more than the average sprinkler, which is only justifiable when assurance can be made that these products will perform as intended. Therefore, on/off sprinkler use in heritage facilities should remain limited.

Selection of specific sprinklers is based on: risk characteristics, ambient room temperature, desired response time, hazard criticality and aesthetic factors. Several sprinkler types may be used in a heritage facility.

All sprinkler systems require a reliable water source. In urban areas, a piped public service is the most common supply, while rural areas generally utilize private tanks, reservoirs, lakes, or rivers. Where a high degree of reliability is desired, or a single source is undependable, multiple supplies may be utilized.

Basic water source criteria include:

- The source must be available at all times. Fires can happen at any time and therefore, the water supply must be in a constant state of readiness. Supplies must be evaluated for resistance to pipe failure, pressure loss, droughts, and other issues that may impact availability.
- The system must provide adequate sprinkler supply and pressure. A sprinkler system will create a hydraulic demand, in terms of flow and pressure, on the water supply. The supply must be capable of meeting this demand. Otherwise, supplemental components such as a fire pump or standby tank must be added to the system.
- The supply must provide water for the anticipated fire duration. Depending of the fire hazard, suppression may take several minutes to over an hour. The selected source must be capable of providing sprinklers with water until suppression has been achieved.
- The system must provide water for fire department hoses operating in tandem with the sprinkler system. Most fire department procedures involve the use of fire attack hoses to supplement sprinklers. The water supply must be capable of handling this additional demand without adverse impact on sprinkler performance.

Sprinkler water is transported to fire via a system of fixed pipes and fittings. Piping material options include various steel alloys, copper, and fire resistant plastics. Steel is the traditional material with copper and plastics utilized in many sensitive applications. Primary considerations for selection of pipe materials include:

- **Ease of installation.** The easier the material is installed, the less disruption is imposed on the institution's operations and mission. The ability to install a system with the least amount of disturbance is an important consideration, especially in sprinkler retrofit applications where building use will continue during construction.
- **Cost of material versus cost of protected area.** Piping typically represents the greatest single cost item in a sprinkler system. Often there is a temptation to reduce costs by utilizing less expensive piping materials that may be perfectly acceptable in certain instances, i.e. office or commercial environs. However, in heritage applications where the value of contents may be far beyond sprinkler costs, appropriateness of the piping rather than cost should be the deciding factor.
- **Contractor familiarity with materials.** A mistake to be avoided is one in which the contractor and pipe materials have been selected, only to find out that the contractor is inexperienced with the pipe. This can lead to installation difficulties, added expense, and increased failure potential. A contractor must demonstrate familiarity with the desired material before selection.
- **Prefabrication requirements or other installation constraints.** In some instances, such as in fine art vaults, requirements may be imposed to limit the amount of work time in the space. This will often require extensive prefabrication work outside of the work area. Some materials are easily adapted to prefabrication.
- **Material cleanliness.** Some pipe materials are cleaner to install than others. This will reduce the potential for soiling collections, displays, or building finishes during installation. Various materials are also resistant to accumulation in the system water, which could discharge onto collections. Cleanliness of installation and discharge should be a consideration.
- **Labor requirements.** Some pipe materials are heavier or more cumbersome to work with than others. Consequently additional workers are needed to install pipes, which can add to installation costs. If the number of construction workers allowed into the building is a factor, lighter materials may be beneficial.

The benefits and disadvantages of each material should be evaluated prior to selection of pipe materials.

Other major sprinkler system components include:

- **Control valves.** A sprinkler system must be capable of shut down after the fire has been controlled, and for periodic maintenance and modification. In the simplest system a single shutoff valve may be located at the point where the water supply enters the building. In larger buildings the sprinkler system may consist of multiple zones with a control valve for each. Control valves should be located in readily identified locations to assist responded emergency personnel.
- **Alarms.** Alarms alert building occupants and emergency forces when a sprinkler water flow occurs. The simplest alarms are water driven gongs supplied by the sprinkler system. Electrical flow and pressure switches, connected to a building fire alarm system, are more common in large buildings. Alarms are also provided to alert building management when a sprinkler valve is closed.
- **Drain and test connections.** Most sprinkler systems have provisions to drain pipes during system maintenance. Drains should be properly installed to remove all water from the sprinkler system, and prevent water from leakage onto protected spaces, when piping service is necessary. It is advisable to install drains at a remote location from the supply, thereby permitting effective system flushing to remove debris. Test connections are usually provided to simulate the flow of a sprinkler, thereby verifying the working condition of alarms. Test connections should be operated every 6 months.
- **Specialty valves.** Drypipe and preaction sprinkler systems require complex, special control valves that are designed to hold water from the system piping until needed. These control valves also include air pressure maintenance equipment and emergency operation/release systems.
- **Fire Hose Connections.** Fire fighters will often supplement sprinkler systems with hoses. Firefighting tasks are enhanced by installing hose connections to sprinkler system piping. The additional water demand imposed by these hoses must be factored into the overall sprinkler design in order to prevent adverse system performance.

System Types

There are three basic types of sprinkler systems: wet pipe, dry pipe and preaction, with each having applicability, depending on a variety of conditions such as potential fire severity, anticipated fire growth rates, content water sensitivity, ambient conditions, and desired response. In large multifunction facilities, such as a major museum or library, two or more system types may be employed.

Wet pipe systems are the most common sprinkler system. As the name implies, a wet pipe system is one in which water is constantly maintained within the sprinkler piping. When a sprinkler activates this water is immediately discharged onto the fire. Wet pipe system advantages include:

- **System simplicity and reliability.** Wet pipe sprinkler systems have the least number of components and therefore, the lowest number of items to malfunction. This produces unexcelled reliability, which is important since sprinklers may be asked to sit in waiting for many years before they are needed. This simplicity aspect also becomes important in facilities where system maintenance may not be performed with the desired frequency.
- **Relative low installation and maintenance expense.** Due to their overall simplicity, wet pipe sprinklers require the least amount of installation time and capital. Maintenance cost savings are also realized since less service time is generally required, compared to other system types. These savings become important when maintenance budgets are shrinking.
- **Ease of modification.** Heritage institutions are often dynamic with respect to exhibition and operation spaces. Wet pipe systems are advantageous since modifications involve shutting down the water supply, draining pipes, and making alterations. Following the work, the system is pressure tested and restored. Additional work for detection and special control equipment is avoided, which again saves time and expense.
- **Short term down time following a fire.** Wet pipe sprinkler systems require the least amount of effort to restore. In most instances, sprinkler protection is reinstated by replacing the fused sprinklers and turning the water supply back on. Preaction and drypipe systems may require additional effort to reset control equipment.

The main disadvantage of these systems is that they are not suited for subfreezing environments. There also may be concern where piping is subject to severe impact damage, such as some warehouses.

The advantages of wet systems make them highly desirable for use in most heritage applications, and with limited exception, they represent the system of choice for museum, library and historic building protection.

The next system type, a dry pipe sprinkler system, is one in which pipes are filled with pressurized air or nitrogen, rather than water. This air holds a remote valve, known as a dry pipe valve, in a closed position. The drypipe valve is located in a heated area and prevents water from entering the pipe until a fire causes one or more sprinklers to operate. Once this happens, the air escapes and the dry pipe valve releases. Water then enters the pipe, flowing through open sprinklers onto the fire.

The main advantage of dry pipe sprinkler systems is their ability to provide automatic protection in spaces where freezing is possible. Typical dry pipe installations include unheated warehouses and attics, outside exposed loading docks and within commercial freezers.

Many heritage managers view dry pipe sprinklers as advantageous for protection of collections and other water sensitive areas, with a perceived benefit that a physically damaged wet pipe system will leak while dry pipe systems will not. In these situations, however, dry pipe systems will generally not offer any advantage over wet pipe systems. Should impact damage happen, there will only be a mild discharge delay, i.e. 1 minute, while air in the piping is released before water flow.

Dry pipe systems have some disadvantages that must be evaluated before selecting this equipment. These include:

- **Increased complexity.** Dry pipe systems require additional control equipment and air pressure supply components, which increases system complexity. Without proper maintenance this equipment may be less reliable than a comparable wet pipe system.
- **Higher installation and maintenance costs.** The added complexity impacts the overall drypipe installation cost. This complexity also increases maintenance expenditure, primarily due to added service labor costs.
- **Lower design flexibility.** There are strict requirements regarding the maximum permitted size (typically 750 gallons) of individual drypipe systems. These limitations may impact the ability of an owner to make system additions.

- **Increased fire response time.** Up to 60 seconds may pass from the time a sprinkler opens until water is discharged onto the fire. This will delay fire extinguishing actions, which may produce increased content damage.
- **Increased corrosion potential.** Following operation, drypipe sprinkler systems must be completely drained and dried. Otherwise, remaining water may cause pipe corrosion and premature failure. This is not a problem with wet pipe systems where water is constantly maintained in piping.

With the exception of unheated building spaces and freezer rooms, dry pipe systems do not offer any significant advantages over wet pipe systems and their use in heritage buildings is generally not recommended.

The third sprinkler system type, preaction, employs the basic concept of a dry pipe system in that water is not normally contained within the pipes. The difference, however, is that water is held from piping by an electrically operated valve, known as a preaction valve. The operation of this valve is controlled by independent flame, heat, or smoke detection. Two separate events must happen to initiate sprinkler discharge. First, the detection system must identify a developing fire and then open the preaction valve. This allows water to flow into system piping, which effectively creates a wet pipe sprinkler system. Second, individual sprinkler heads must release to permit water flow onto the fire.

In some instances, the preaction system may be set up with an interlock feature in which pressurized air or nitrogen is added to system piping. The purpose of this feature is twofold: first to monitor piping for leaks and second to hold water from system piping in the event of inadvertent detector operation. The most common application for this system type is in freezer warehouses.

The primary advantage of a preaction system is the dual action required for water release: the preaction valve must operate and sprinkler heads must fuse. This provides an added level of protection against inadvertent discharge, and for this reason, these systems are frequently employed in water sensitive environments such as archival vaults, fine art storage rooms, rare book libraries and computer centers.

There are some disadvantages to preaction systems. These include:

- **Higher installation and maintenance costs.** Preaction systems are more complex with several additional components, notably a fire detection system. This adds to the overall system cost.

- **Modification difficulties.** As with drypipe systems, preaction sprinkler systems have specific size limitations which may impact future system modifications. In addition, system modifications must incorporate changes to the fire detection and control system to ensure proper operation.
- **Potential decreased reliability.** The higher level of complexity associated with preaction systems creates an increased chance that something may not work when needed. Regular maintenance is essential to ensure reliability. Therefore, if the facility's management decides to install preaction sprinkler protection, they must remain committed to installing the highest quality equipment, and to maintaining these systems as required by manufacturer's recommendations.

Provided the application is appropriate, preaction systems have a place in heritage buildings, especially in water sensitive spaces.

A slight variation of preaction sprinklers is the deluge system, which is basically a preaction system using open sprinklers. Operation of the fire detection system releases a deluge valve, which in turn produces immediate water flow through all sprinklers in a given area. Typical deluge systems applications are found in specialized industrial situations, i.e., aircraft hangers and chemical plants, where high velocity suppression is necessary to prevent fire spread. Use of deluge systems in heritage facilities is rare and typically not recommended.

Another preaction system variation is the on/off system which utilizes the basic arrangement of a preaction system, with the addition of a thermal detector and nonlatching alarm panel. The system functions similar to any other preaction sprinkler system, except that as the fire is extinguished, a thermal device cools to allow the control panel to shut off water flow. If the fire should reignite, the system will turn back on. In certain applications on/off systems can be effective. Care, however, must be exercised when selecting this equipment to ensure that it functions as desired. In most urban areas, it is likely that the fire department will arrive before the system has shut itself down, thereby defeating any actual benefits.

Sprinkler Concerns

Several common misconceptions about sprinkler systems exist. Consequently, heritage building owners and operators are often reluctant to provide this protection, especially for collections storage and other water sensitive spaces. Typical misunderstandings include:

- *When one sprinkler operates, all will activate.* With the exception of deluge systems (discussed later in this leaflet), only those sprinklers in direct contact with the fire's heat will react. Statistically, approximately 61% of all sprinkler controlled fires are stopped by two or less sprinklers.
- *Sprinklers operate when exposed to smoke.* Sprinklers function by thermal impact against their sensing elements. The presence of smoke alone will not cause activation without high heat.
- *Sprinkler systems are prone to leakage or inadvertent operation.* Insurance statistics indicate a failure rate of approximately 1 head failure per 16,000,000 sprinklers installed per year. Sprinkler components and systems are among the most tested systems in an average building. Failure of a proper system is very remote. Where failures do occur, they are usually the result of improper design, installation, or maintenance. Therefore, to avoid problems, the institution should carefully select those who will be responsible for the installation and be committed to proper system maintenance.
- *Sprinkler activation will cause excessive water damage to contents and structure.* Water damage will occur when a sprinkler activates. This issue becomes relative, however, when compared to alternative suppression methods. The typical sprinklerwill discharge approximately 25 gallons per minute (GPM) while the typical fire department hose delivers 100–250 GPM. Sprinklers are significantly less damaging than hoses. Since sprinklers usually operate before the fire becomes large, the overall water quantity required for control is lower than situations where the fire continues to increase until firefighters arrive.

One final point to consider is that the water damage is usually capable of repair and restoration. Burned out contents, however, are often beyond mend.

- *Sprinkler systems look bad and will harm the building's appearance.* This concern has usually resulted from someone who has observed a less than ideal appearing system, and admittedly there are some poorly designed systems out there. Sprinkler systems can be designed and installed with almost no aesthetic impact.

To ensure proper design, the institution and design team should take an active role in the selection of visible components. Sprinkler piping should be placed, either concealed or in a decorative arrangement, to minimize visual impact. Only sprinklers with high quality finishes should be used. Often sprinkler manufacturers will use customer provided paints to match finish colors, while maintaining the sprinkler's listing. The selected sprinkler contractor must understand the role of aesthetics.

To help ensure overall success, the sprinkler system designer should understand the institution's protection objectives, operations, and fire risks. This individual should be knowledgeable about system requirements and flexible to implement unique, thought-out solutions for those areas where special aesthetic or operations concerns exist. The designer should be experienced in the design of systems in architecturally sensitive applications.

Ideally, the sprinkler contractor should be experienced working in heritage applications. However, an option is to select a contractor experienced in water sensitive applications such as telecommunications, pharmaceuticals, clean rooms, or high tech manufacturing. Companies including AT&T, Bristol Meyers Squibb, and IBM have very stringent sprinkler installation requirements. If a sprinkler contractor has demonstrated success with these types of organizations, then they will be capable of performing satisfactorily in a heritage site.

The selected sprinkler components should be provided by a reputable manufacturer, experienced in special, water sensitive hazards. The cost differential between average and the highest quality components is minimal. The long term benefit, however, is substantial. When considering the value of a facility and its contents, the extra investment is worthwhile.

With proper attention to selection, design, and maintenance, sprinkler systems will serve the institution without adverse impact. If the institution or design team does not possess the experience to ensure the system is proper, a fire protection engineer experienced in heritage applications can be a great advantage.

Water Mist

One of the most promising automatic extinguishing technologies is the recently available fine water droplet, or mist systems. This technology represents another tool that can provide automatic fire suppression in some cultural property applications. Potential uses include locations where reliable water supplies do not exist, where even sprinkler water discharges are too high, or where building construction and aesthetics impact the use of standard sprinkler pipe dimensions. Mist systems may also be an appropriate solution to the protection void left by the environmental concerns, and subsequent demise, of Halon 1301 gas.

Mist technology was originally developed for offshore uses such as on board ships and oil drilling platforms. For both of these applications, there is a need to control severe fires while limiting the amount of extinguishing water, which could impact vessel stability. These systems have been extensively approved by a number of domestic and international marine organizations, and have been a protection standard for the past 8–10 years. They have a solid track record dealing with maritime fires. These systems have also been used in several land based applications, and have a number of listings, primarily in Europe, where their effectiveness has been recognized. Some systems have recently received approvals for North American land based uses.

Mist systems discharge limited water quantities at higher pressures than sprinkler systems. These pressures range from approximately 100 to 1,000 psi, with the higher pressure systems generally producing larger volumes of fine sprays. The produced droplets are usually in the 50 to 200 micron diameter range (compared to 600–1,000 microns for standard sprinklers), resulting in exceptionally high efficiency cooling and fire control, with significantly little water. In most situations, fires are controlled with approximately 10-25% of the water normally associated with sprinklers. Water saturation that is often associated with standard firefighting procedures is decreased. Other benefits include lower aesthetic impact and known environmental safety.

Typical water mist systems consist of the following components:

- **Water supply:** Water for a system may be provided by either the piped building system or a dedicated tank arrangement. In some instances, lower pressure systems may use existing sprinkler piping. For most, however, supplemental pumps will be required. Other options include dedicated water/nitrogen storage cylinders, which can deliver a limited duration supply.
- **Piping and nozzles:** Piping can be greatly reduced when compared to sprinklers. For low pressure systems, pipes are generally 25-50% smaller than comparable sprinkler piping. For high pressure systems, piping is even smaller with the 0.50-0.75 inch diameters as the norm. Like sprinklers, nozzles are individually activated by the fire's heat, and are selected to cover a certain size hazard. Their sizes are comparable to a low profile sprinkler.

- **Detection and control equipment:** In some instances, mist discharge can be controlled by selected, high reliability intelligent detectors or by an advanced technology VESDA smoke detection system. These systems represent the premier, state of the art, fire detection technology that can provide very early warning of a developing fire, as well as reduce the probability of inadvertent discharge.

At this point, one of the main drawbacks to mist systems is their higher cost, which can be 50–100% greater than standard sprinklers. This cost, however, may be reduced due to possible installation labor savings. In rural applications, where reliable sprinkler water supplies can be expensive, mist systems may be comparable or less than standard sprinklers. Another problem is that these systems do not have the variety of approvals and listings commonly associated with sprinklers. As such, they may not be as recognized by fire and building authorities. In addition, the number of contractors who are familiar with the technology is limited. These concerns are diminishing, however, as use of these systems becomes more widespread.

Planning

Before planning can proceed at any administrative level, the objectives have to be defined in measurable or tangible terms and related to the policy laid down by the executive.

For each administrative level, this will permit the evaluation of the degree of effort which must be expended to attain the objectives. The comparison of the present efficiency of the fire control effort in a field administrative unit area to the defined objectives, will assist in determining the elapsed time standards for each of the major segments of the fire control job, i.e., detection, communication, transportation, and suppression.

This will provide the basis for the fire control planning job.

The translation of the fire control policy, as stated by the executive, into measurable or tangible objectives, which can be used as outlined above, is not always a simple task.

For a given administrative area, the definition of the objectives will vary according to land use, and the values at stake, present and potential.

Special values will also occur such as human life, man-made improvements, parks and research reserves. The fire control objectives acceptable to the economy of a given area will depend on factors such as the present and potential wood production requirements from the area by the industry, the growth and reproductive characteristics of the forest, the character of the soil and site, accessibility, potential losses from insects and disease and control costs, and the cost of a given degree of fire control intensity.

In developing realistic objectives for the given area, the approach which we favour is to develop a fire control oganization whose level of efficiency is sufficient to accomplish the following:

1. During a normal or average fire season to place under control within 24 hours (of ignition) all fires which occur, despite preventive measures, and to limit the maximum area burned per fire to 50 acres or less.
2. During an abnormal or bad fire season to be able to expand quickly enough to place most fires which occur, despite preventive measures, under control within 24 hours (of ignition), and/or to limit the maximum acreage burned per fire to 500 acres.

(Except that the largest permissible fire for special areas will be as follows: plantations-50 acres; parks-5 acres; research reserves - 0.05 acres; etcetera.)

Before detailed presuppression planning can be undertaken, certain basic data must be obtained. These will include values at stake, fire risk, fuel zones, and other factors of fire danger.

The strategy and tactics required to realize the fire protection objectives are determined by the combined effect of the fire control factors.

Constant (Relatively)

1. Values at stake;
2. Fire Risk;
3. Fuel Types;
4. Topography.

Variable

5. Fuel inflammability due to weather factors;
6. Wind velocity;
7. Temporary fire risk.

Besides the factors which are listed above, the fire control job is also determined by the efficiency of the fire control organization which is determined by its planning ability, the action taken, the capabilities of the manpower involved, and technological progress.

The organizational factors are divided into the following groups:

1. Prevention;
2. Detection;
3. Communication;
4. Transportation;
5. Availability and skill of manpower;
6. Availability and type of equipment.

The measurement, weighing and correlation of all of these factors will indicate the fire load at a given time for a given area and the strategy and tactics required to attain the objectives.

Value at Stake Study

The forest fire control objectives for a given area should be based on its economic, social, and developmental requirements and potentialities.

Land use planning would establish the boundaries of the forest zone, the agricultural zone, and any other zones outlining areas of similar human, industrial, aesthetic or other values.

The values at stake within a given area are shown on a map as follows:

- Zone One-Extreme Values at Stake: Human lives, permanent improvements, parks, research areas, plantations.
- Zone Two-High Values at Stake: Forest lands under intensive forest management, forest lands supplying present or future industries and markets, planted areas, some watershed and recreational areas, specific wildlife habitat.
- Zone Three-Medium Values at Stake: Undeveloped forest lands which will become accessible and placed under management in the foreseeable future.
- Zone Four-Low Values at Stake: Land which will not be accessible in the foreseeable future.

Fire Risk Study

Fire risk is defined as the relative chance or probability of fire starting as determined by the presence or absence of causative agencies.

The causative agencies divide into two main classes:

1. Fires started by man.
2. Fires started by natural phenomena, e.g. lightning.

Man-caused fire risk is not static over long periods in an area, but changes in accordance with the activities of man. The risk of lightning fires varies by years but certain physiographic regions average a higher relative incidence over the years.

The tracks of lightning storms should be plotted as accurately as field data permit, as a basis for determining the fire detection patrol requirements from day to day.

Fire risk can be divided into two main phases, one on the basis of the fire occurrence history on given areas over a relatively long time period, and the second on the basis of the fire occurrence history on given areas over a relatively short time period.

A study of the first phase for the given area will provide the planner with the basic data required to determine the fire occurrence zone to which the area belongs.

A study of the latter will provide the planner with data required in organizing fire control action for a given period and will assist in deciding on the peak fire load probabilities.

The decision on the peak fire occurrence load probnbilities over a short period of time is critical to the success or failure of the fire control organization.

This determination is based on past experience and future probabilities.

In order to derive the potential fire load, a picture of the probable fire risk by zones of relative intensity is required for each planning unit.

The best way to do this is to prepare fire occurrence maps, for a given period in the recent past, showing the starting point of each fire with an appropriate symbol to indicate the size-class, cause, and year.

From the occurrence maps, occurrence zone maps showing the relative intensity of fire occurrence may be made. These may combine man-caused and lightning fires, or separate them where necessary.

The number of fires per 1,000 acres for 10 years is suggested as a means of measuring the intensity index.

The forest fires which occurred during the given period are to be plotted, according to a standard legend, on a map having a scale of 2 miles = 1 inch. Then the fire occurrence zones are plotted on xylonite or other transparent material, gridded into 1,000 acre squares, which is placed over the map.

The intensity index defining each fire occurrence zone will be based on the number of fires per 1,000 acres for the stated 10 year period according to an index scale.

Zones of Relative Fire Risk

(Give occurrence index for each zone if possible)

Zone One-Extreme Risk: Highest occurrence intensity in the past with similar probability for the next period.

Zone Two-High Risk: High occurrence intensity in the past with similar probability for the next period.

Zone Three-Medium Risk: Medium occurrence intensity in the past with similar probability for the next period.

Zone Four-Low Risk: Fire occurrence in past uncommon with similar probabilities for the next period.

Fule hazard Zone Study

In planning overall fire control strategy for a given area a fuel hazard zone map is required as an important factor to be combined with the maps showing the zones of values at stake and fire risk. The fuel hazard zone map is derived from the fuel type maps, but if these aren't available then from the best available data recording forest types, cutover, insect infestation, and blowdown data for the area concerned.

Fuel Hazard Zones

Zone One-Extreme Fuel Hazard: Continuous fine, medium and heavy fuels of extreme volume exposed to wind and sun, e.g. coniferous slash between 6 months and 4 years old and insect-killed stands. (extreme rate of spread and control difficulty)

Zone Two-High Fuel Hazard: Continuous to patchy fine, medium and heavy fuels of high volume, e.g. coniferous slash, insect-killed stands, some old burn areas and some coniferous timber stands. (high rate of spread and control difficulty)

Zone Three-Medium Fuel Hazard: Continuous to patchy or broken fuels, mostly green, of medium arrangement from standpoint of inflammability, e.g. mixed growing timber stands. (medium rate of spread and control difficulty)

Zone Four-Low Fuel Hazard: Patchy or broken green fuels of poor arrangement from standpoint of inflammability, e.g., tolerant hardwood stands. (low rate of spread and control difficulty)

Fire Danger Zone Study

The maps showing the values to be protected, fire risk zones, and fuel hazard zones, should be combined into a map showing the zones of fire danger.

This fire danger zone map is used in conjunction with the fire control objectives and elapsed time standards, expected fire incidence, the accessibility map, the burning index rating system and visibility conditions, in determining the potential fire load for the given area and time and thus the general strategy and tactics of the fire control organization for the following:- prevention, detection, communications, transportation, location and strength of manpower, equipment requirements by location and suppression plans.

The map showing the zones of fire danger might define the zones as follows:

Fire Danger Zones

Zone One-Extreme Fire Danger: Includes areas of extreme fuel hazard, extreme to high fire risk and extreme to high values at stake.

Zone Two-High Fire Danger: Includes areas of high fuel hazard, high fire risk and high values at stake.

Zone Three-Medium Fire Danger: Includes areas of medium fuel hazard, medium fire risk and medium values at stake.

Zone Four-Low Fire Danger: Includes areas of low fuel hazard, low fie risk and medium to low values at stake.

Burning Index and Visibility

The main variable factors include the inflammability of the fuel types as a result of weather conditions and season of year, wind velocity, temporary fire risk and visibility conditions.

The measurement of the fuel inflammability and burning conditions must be done scientifically by a standard method to permit comparison from area to area and season to season.

To realize this, standard instructions for taking the necessary measurements and correlating them is necessary.

Visibility conditions affect the detection job and the current situation must be known to the fire control officers for use in the organization of the fire detection job to meet the elapsed time objective.

Particularly critical is the determination of the areas requiring aircraft and ground detection patrols and their timing and frequency.

Special patrol maps based on different classes of visibility from the fixed detection system are valuable in despatching such patrols.

Accessibility Study (Travel Time)

The accessibility study will evaluate the transportation situation relative to the elapsed time objectives as they affect the location and transportation of the fire suppression forces to the fire in a given area.

To compile the data in usable form, a travel time map is first made up.

This map is used as the base for showing the actual travel time required, using the most practical means of transportation, from a given location at which suppression crews are located to the various points in the area covered by each location or headquarters.

It can then be determined whether additional placement of crews and/or other transportation techniques are required in order to meet the elapsed time objective between getaway and reaching the fire by the suppression forces.

If the elapsed time objectives for the given area cannot be realized then methods to retard the rate of spread after detection are necessary.

An example of such a method is aerial water dropping.

OCCURRENCE OF PEAK FIRE LOADS

A peak fire load can occur when there is an extraordinary combination of the following circumstances:

1. The fire prevention effort fails to prevent more man-caused forest fires from igniting than the fire control establishment can suppress in the time and with the efficiency necessary to prevent the fire load from increasing beyond the maximum capability of the organization.
2. Lightning fires occurrence is above that expected and/or in relatively inaccessible places.
3. The burning index is above normal and/or combined with a drought and/or certain wind velocities and/or hazardous fuel conditions.

4. The detection of each fire is delayed beyond the elapsed time objectives due to human error and/or inadequate detection coverage due to poor visibility conditions and lack of sufficient aerial patrols.

The moment of truth for a fire control organization is its adequacy in meeting a peak fire load.

Fire suppression involves the application of the proper tools and materials in the right way, in the right amount, at the right time, in accordance with the factors involved, by fully organized and adequately trained men.

Due to the economics involved, unit fire control organizations are geared to meet the fire control objectives during "normal" fire control loads for the given area.

However, the plans to meet a "normal" fire control load should include, of necessity, a method for rapid build-up to meet "above normal" or peak fire load conditions.

Adequate plans to meet a peak fire load must include the following:

1. The prompt recognition of the build-up and potential peak load situation.
2. An adequate fire detection system for each degree of visibility.
3. A fire control organization of sufficient flexibility to permit movement of fire control power into peak fire load areas.

If the fire control plans are adequate and have been implemented to the necessary degree, then the assessment and measurement of the factors involved in producing a peak fire load is continuous, and prompt recognition of a build- up should be possible, subject to the coverage provided by the assessment and measurement organization.

An adequate detection system depends on the experience and know-how of the personnel involved coupled with their use of every possible detection aid.

During a peak fire load, with its usual requirement of providing fire control power above that which the normal force involved can supply, the adequacy of the movement of man-power, equipment, and supplies into the areas involved will depend also on the adequacy of the pre-suppression planning and on the ability of the despatchers concerned with larger unit areas, e.g., regions and province.

Besides the despatching necessities, such as an adequate communication system, actual experience during extraordinary peak load conditions has proven the value to the despatching officers of summing up some of the important fire control factors on a map of the area involved.

Such a map would be prepared once or twice a day or even oftener, if necessary and possible, from observations reported by towers, weather stations, aircraft and co-operators. The observations recorded on the map would include the location and movement of lightning storms, precipitation total within the last 24 hours and the total for the last 7 days.

With the addition of the known data on the other fire control factors, e.g.-human fire risk and visibility, the zones of probable fire load are drawn.

This map is then used to plan, for the next time period, the strategy of fire detection and the requirements and distribution of man-power, equipment and supplies over administrative areas and the province.

Organization

In the United States, there is broad recognition that current wildfire management practices must fundamentally change. Whether this is due to concerns about climate change, rising suppression costs, or the growing number of values at risk, research consistently indicates that US fire management is a wicked policy problem and one that, no matter how it is defined, is getting worse with growing negative effects to society. Multiple scholars also have noted that fire governance must be viewed as a complex social-ecological system, and actors must embrace a diversity of solutions over time and in different places.

An overview of US fire management problems and solutions

Over the last several decades, fire activity has increased in the United States, particularly the western United States (hereinafter, "the West"), with significant and increasing costs to society. Fire responders continue to experience fatalities every year with no evidence of reductions in fatalities relative to exposure, to our knowledge (NIFC 2016). In addition, there are a number of civilian fatalities every year, although trends are unknown as there is no national database consistently tracking civilian wildfire fatalities. The expanding population base, both in and adjacent to fire-prone areas, further increases the social values at risk, and the challenges of fire management will only get worse as fire seasons lengthen with climate change. Yet, despite increases in fire activity, many forested areas in the West face a fire deficit, with an accumulation of areas in need of treatment and entire fire-adapted landscapes likely becoming "endangered". Recommendations boil down to the need for more ecologically beneficial fires and fewer fires with negative ecological or social impacts.

In terms of federal fire management, there is broad recognition that the historical and ongoing emphasis on full suppression is problematic, leading to an unnatural accumulation of fuels, and that short-term fire suppression success creates a fire debt that eventually comes due. Many fire-adapted ecosystems would benefit from low-moderate severity fire to maintain or restore ecological integrity. However, under current management paradigms, US fire managers are unlikely to be able to maintain, much less increase, the acreage that is treated with fire prepared for fire. Roughly 98% of fires are still actively suppressed and rapidly contained (USFS 2015). Some argue that the Forest Service has poorly capitalized on opportunities to manage wildland fires for resource benefit, instead perpetuating a model of fire exclusion, which can lead to ever worsening feedbacks. In addition, the amount of money spent on fighting wildfire on national forests, both in absolute terms and as a proportion of the US Forest Service's budget, has increased dramatically since 2000 (USFS 2015). As of 2017, wildfire management accounted for about 60% of annual appropriations for the agency, leaving less money for everything else—perhaps most notably the restoration work that is designed to support more resilient and fire-adapted landscapes that would reduce the need for fire suppression in the future (USFS 2015). The Consolidated Appropriations Act of 2018 included a "fire funding fix," scheduled to go into effect in 2020, which would increase the available funds for wildfire management and thereby decrease the percentage of the agency's budget that goes to wildfire management. Nonetheless, if the status-quo of suppressing almost all wildfires is maintained, management practices may do little to alleviate, and may exacerbate, fire problems.

The extent of the problem in the United States is not lost on communities, land and fire managers, policy makers, or scientists. The scientific literature articulates the need to adapt to living with more fire, reduce exposure of responders and civilians, manage costs more effectively, increase restoration activities, and improve the ability to use fire as a management tool. The research also indicates that communities in the West increasingly understand the importance of fire in the landscapes where they reside. Land managers in surveys also reflect an understanding of and commitment to these interrelated objectives. And over time, as we discuss in subsequent sections of the paper, the US government has articulated the need for fire-adapted human communities, forest restoration, and improved approaches to suppression as exemplified through acts of Congress and numerous agency policies and actions.

In summary, there is broad agreement that current fire management practices are leading to undesirable outcomes and are unsustainable in the long run. It has become starkly apparent that recognizing that there is a problem is one thing, while solving it is entirely another. Reforming current fire management approaches will require work across social, economic, ecological, and organizational variables. Significant efforts need to occur on the social and economic pieces of the puzzle so that communities can become more fire adapted, and so that more forest restoration activities can take place where they are deemed appropriate. More work is also needed so that fire responders can enhance the safety and effectiveness of fire management actions, including through capitalizing on opportunities to expand the footprint of fires to achieve long-term land and resource objectives

Examining Forest Service institutions for fire and forest management

The Forest Service is made up of five primary branches, two of which are directly involved in fire management: the National Forest System (NFS) and State and Private Forestry. The NFS manages the national forests and is responsible for resource distribution for land management, land and resource management planning (*i.e.*, forest planning), project-level planning, and project implementation to meet forest plan goals. State and Private Forestry, in addition to engaging in landowner and state-level partnerships and assistance programs, houses the Fire and Aviation Management (FAM) program, which oversees fire prevention, pre-suppression preparedness, and suppression activities within the agency. Although many fire staff members are NFS employees, their funding, fire training, planning, and response operations typically follow FAM guidance and requirements. Historically, FAM and NFS have shared responsibility for prioritizing investments in vegetation treatments to restore forest conditions and reduce hazardous fuel loads, although budget controls for these activities shifted primarily to NFS in fiscal year (FY) 2018.

Supporting a more integrated approach

The complexity of today's public management challenges requires greater coordination within and across organizations. While some specialization within the Forest Service of its wildland fire and land management branches may be necessary, better connections and communication between the two sides of the agency could improve the ability to more effectively navigate goal ambiguity. In addition, improved integration of planning exercises and more clarity on how to prioritize among goals may be in order. In this section, we offer our suggestions for paths forward, constraining our recommendations to actions that could be undertaken within the current legal framework and organizational structure of the Forest Service.

As noted earlier, eliminating goal ambiguity is not desirable or feasible for Forest Service land and fire management. However, in the face of goal ambiguity and competing problem definitions, it can be useful to develop a "meta-frame," which is a framing of a problem that incorporates multiple problem definitions. A shared meta-frame could involve different constituencies and actors in "constructing a shared narrative that recognizes multiple voices, teases out the implications of these value preferences, and seeks to resolve conflicts". This presents an area of opportunity for improved internal dialogue with the agency, as a starting point, perhaps towards defining fire management as a complex risk management problem. In addition, developing a meta-frame requires dialogue among multiple constituencies. For fire management, this kind of stakeholder dialogue is important because of the emphasis on collaboration in policy, the role of local partners and stakeholders in preparing and responding to fire, and because local collaborative contexts will have a significant impact on implementation of ambiguous policy goals. Working with stakeholders (*e.g.*, community representatives, partner agencies, non-governmental organizations that bring capacity, etc.) to develop a meta-frame could facilitate effective response to complex problems by allowing for improved understanding of the nature of the problem from multiple perspectives and identification of a wider range of possible solutions, along with support for their implementation. External partners also can offer a degree of accountability to long-term goals and bring support for those goals to fire events, potentially alleviating some of the political pressure and natural decision biases that lead to a relatively greater emphasis on managing for short-term risks in the face of goal ambiguity. Long-term collaboration and larger-scale planning under approaches like CFLRP and other Forest Service restoration approaches likely support meta-framing and a possible a path forward; internal agency and stakeholder collaboration during forest plan revisions, which could embrace a greater emphasis on fire as a natural process critical for long-term sustainability, also may present an opportunity.

Another step the agency could take to support the long-term goals of reintroducing more fire into forested ecosystems and minimize the risks associated with not doing so, would be to decrease barriers to engaging in strategic and transparent fire management planning, which would allow for more upfront consideration of various tradeoffs. In particular, long-term versus short-term considerations can be undertaken deliberatively and over time, outside the context of emergency management. We support current agency efforts in this direction, including creation of operationally relevant fire plans that consider factors related to fire control opportunities and firefighter safety. We also support the notion of developing plans that are spatially and temporally scalable, and that are designed to be flexible and adaptive. A related opportunity lies in better integrating fire management planning into forest and project-level planning so that tradeoffs can be evaluated at the plan stage and built into project-level planning more effectively. During the planning process, identification could be made of areas where fire might be managed on the land to meet different management objectives, where it would likely need to be suppressed, and where to prioritize fuels treatment. Some of this has happened in the past but often outside of the NEPA process, limiting opportunities for transparency, deliberation, and stakeholder involvement.

In essence, we suggest that incorporating conversations about fire management more regularly into formal decision-making processes could lay valuable groundwork for both forest and project planning, while engaging all participants in creating necessary meta-frames for fire management. Ideally, pre-season discussions about potential strategies and tradeoffs with different resource personnel or other agency partners could expand the decision space during fire events. Such activities could be used as a means of setting expectations for response for both an IMT as well as the community, with the goal of ensuring that effective plans for fire management are in place that can support more burning and have stakeholder support. Over time, this may create more pressure to justify suppression decisions when they compromise long-term management goals. Importantly, these activities require that the agency has adequate capacity, so that personnel trained in both fire management and fire ecology can participate in planning processes and have the time and skills to effectively engage in collaboration with partners and stakeholders.

Another opportunity to improve decision making is to build greater accountability into the incident command process, which has been a historically difficult area for the public to gain any oversight or involvement. Accountability is central to successful public administration and can augment understanding of where current practices are contributing to the status quo. Organizations with activities subject to direct observation also are better at evaluating progress towards goals. Although there already are multiple avenues for public involvement and oversight within the NFS's planning processes, the incident response side of the equation may be an area in which greater accountability could improve outcomes. One core area would be evaluating the quality of line officer and incident commander decisions on large fire incidents. We recognize that line officers and incident commanders may see greater accountability measures as potentially taking time away from more important activities or second-guessing their expertise in what are high-pressure situations. However, the time is ripe to explore opportunities to introduce greater accountability procedures formally, particularly given the current emphasis from Congress on cost accountability. For example, tracking variation in suppression resource use and identifying high-use IMTs could support greater accountability around cost containment and firefighter exposure. An important aspect of this would be to evaluate patterns of behavior over time, as opposed to confining evaluation to isolated fire assignments or even fire seasons.

Finally, we segue from accountability to the related problem of performance measurement and how this may be influencing decision making in the face of goal ambiguity. Ongoing evaluation of performance measures is always essential to ensure that they are incentivizing the desired mix of activities and meeting communication needs acknowledged a challenge in performance measurement for fire risk reduction a decade ago, writing, "Current performance indicators such as acres treated may not be the best proxy for assessing whether long-term risk is mitigated" (Steelman and Burke 2007: 70); they pointed to this as a factor for why fire suppression continues to be emphasized over management of natural fire, despite policy change. One area for which better measures may be needed is in tracking the ecological outcomes associated with restoration activities, including the reintroduction of fire. In the face of goal ambiguity, because decision makers will favor measurable outcomes, it is important to consider how to better measure ecological outcomes, both to create strong incentives to manage fire for long-term ecological benefits and to effectively communicate success to policy makers. One opportunity may be to keep acres-treated targets, but to also develop one-, five-, and ten-year goals for priority watersheds, communities, or ecosystems at risk. Short-term measures could be aggregated into more complex multi-year measures to communicate success over time.

Another possibility for enhanced performance measurement would be to move away from strictly outcome-based reporting of beneficial burned acres (*i.e.*, fire effects judged to have been favorable in achieving forest plan objectives independent of the response strategy), to a system that more strongly couples response decisions with reporting accomplishments. The goal would be to provide clearer incentives during incidents to manage natural fires for something other than suppression, in areas where forest plans or project decisions allow for fire. We are not advocating against ecological monitoring to determine post-fire conditions in relation to desired conditions but, rather, suggesting that claiming such acres as "accomplishments" might require a stronger demonstration of alignment between pre-fire plans, fire response objectives, and fire outcomes. Such a shift could require, for instance, the agency to more clearly articulate how much effort put into preventing an acre from burning is tolerable for that same acre, if it does ultimately burn, to be claimed as an accomplishment. In the ideal, such systemic changes could better align objectives, strategy, and performance by evaluating outcomes in relation to how management decisions and actions influenced attainment of land management objectives. A final issue, particularly for FAM, is to provide more dialogue about the import of different performance measures and how to prioritize among them. At present, priorities are unclear and different measures provide conflicting incentives. Further, some aspects of success, as defined by the agency, have no clear, associated measures, making it likely that these activities are under-incentivized, and also making it likely that the agency might miss opportunities for organizational learning and improvement.

Agencies & Organizations

International Association of Wildland Fire	International Association of Wildland Fire The International Association of Wildland Fire (IAWF) is a non-profit, professional association representing members of the global wildland fire community. The purpose of the association is to facilitate communication and provide leadership for the wildland fire community.
	National Wildfire Coordinating Group NWCG is an operational group designed to coordinate programs of the participating wildfire management agencies.
NIFC	The National Interagency Fire Center The National Interagency Fire Center (NIFC), is the nation's support center for wildland firefighting. The NIFC website offers up-to-date information on fires nationwide, contracting, maps, news, and a daily situation report. NIFC also offers a wealth of fire prevention and fuel control information, and training and qualification information.
PREVIEW N/A	Wildland Fire Leadership Council The Wildland Fire Leadership Council (WFLC) was established in April 2002 by the Secretaries of Agriculture and the Interior to provide an intergovernmental committee to support the implementation and coordination of Federal Fire Management Policy. A Memorandum of Understanding (PDF, 1.6 MB) was signed in April 2010 by the Secretary of the Interior, Secretary of Agriculture, and Secretary of Homeland Security to authorize the continuation of the WFLC. The Council meets regularly to provide oversight and coordination of the Federal Wildland Fire Management Policy.
	Forests and Rangelands This website provides fire, fuels, and land management information to government officials, land and fire management professionals, businesses, communities, and other interested organizations and individuals. Particularly, the National Cohesive Wildland Fire Management Strategy (http://www.forestsandrangelands.gov/strategy/index.shtml) is a collaborative process with active involvement of all levels of government and non-governmental organizations. Fifteen listening forums were held across the nation from April through June 2010 where more than 400 participants representing federal, state, tribal, local, and non-governmental organizations discussed their concerns about land management and wildfire related risks. The input from these forums could be found on the website.

FEMA	Federal Emergency Management Agency Wildfire Federal Emergency Management Agency Wildfire.
FEMA	FEMA Ready is a national public service advertising (PSA) campaign designed to educate and empower Americans to prepare for and respond to emergencies including natural and man-made disasters. Its Wildfire page provides information on how to be better prepared for a wildfire.
SWCC	Southwest Coordination Center (SWCC) This website provides four primary components: • Incident Information provides general information pertaining to wildland fire, agency and unit contact numbers, and other relevant information • Logistics Operations provides operational and managerial services specific to aviation, crews, equipment and supplies, and overhead (including IMT's). Predictive Services provides operational products and services to the wildland fire manager. • Fire Management and Administration provides various tools in order to make fire management programs work.

Protecting forests in India from disastrous fires

India's commitment to sustainable development is clearly demonstrated through its innovative and progressive forest policies. The Government's policy of incentivising state governments to improve their forest cover is evident in the 14th Finance Commission's allocation of 7.5% of total revenues on the basis of the state's forest cover. This makes India the implementer of the world's largest Payment for Environmental Services scheme.

Over the last few years, the forest and tree cover in the country has been steadily increasing, and at present, it stands at 24.16% of the total geographic area. This affirms that sustainable forest management and long-term thinking about natural assets are foundations for strong and sustained growth. This is not to say that there are no challenges. Forest fires are a leading cause of forest degradation in India, and the current pattern of widespread and frequent fires could make it more difficult for India to meet its long-term goal of bringing 33% of its geographical area under forest & tree cover and to achieve its international commitment to create additional carbon sinks of 2.5 billion to 3 billion tons worth of CO2 equivalent by 2030.

Forest fire can threaten lives and property. In India, different agencies are mandated with dealing natural disasters including forest fire. Developing institutional mechanisms to improve coordination among various players is a key ingredient for effective management. Forest fire will continue to be managed primarily by state forest departments, but will also need to be a part of disaster planning and the forest department needs to be involved in this process. Because other agencies, such as local police, fire departments and disaster management agencies, may be called in case of large fires, it is important that they are trained in forest fire suppression methods. Moreover, joint trainings should be organized to coordinate between these departments.

In India, local people who depend on forests for their livelihood and to supplement their incomes are the ones who often set fires in the forests. In many cases, they use fire as a tool to obtain the goods and services on which they rely, for example, by burning ground cover to get a fresh flush of grass for consumption by their livestock. Local community members are also the ones who are called to help respond to fires when they spread and burn out of control. While much is already being done at the grassroots level by the forest departments to raise awareness about the impacts of forest fires among local people, increasing the effectiveness of this engagement and encouraging changes in the way people use fire will require providing the right incentives, guided by the overall aim of enhancing the services provided by healthy forests to communities.

At the same time, it is necessary to have comprehensive information on forest fires in India. Forest Fire Information Systems have a role to play in prevention, detection, suppression, and recovery, that is, across the spectrum of needed interventions. Although many states have tried out innovative solutions for fire prevention and management, not all the states are aware of these developments. To leverage this knowledge about what works, it needs to be collected in a systematic way and shared. Systems using satellite monitoring to alert forest managers and the public about active forest fires in their area, pioneered by the Forest Survey of India and states such as Madhya Pradesh, or the early warning system for forest fire danger that is being developed by the Forest Survey of India, are key ingredients of a forest fire management plan. Other states also have these systems, and states can learn from each other's experiences to make these systems more effective.

Moreover, making investments in modern technologies for fire detection, as well as adequate fire suppression and safety equipment, while at the same time working to build capacity across the ranks of forest officers and improving accountability will help the forest department deliver on its mandate.

Techniques of Forest Fire Control

Many people know how to stop, drop, and roll. Some even know how to employ a fire suppression blanket and how to properly operate a fire extinguisher. But for the vast majority of people, fire is something you brag about being able to make out of two sticks, or something you cook dinner on top of, or just something you run away from—certainly not something you want to have to fight to put out. This is all probably for the best—battling a wildfire is no straightforward task. The men and women who do run toward an unexpected blaze have been well trained in strategies that can suppress the flames of a wildfire.

When wildland firefighters are called to action, they immediately assess the situation to figure out how they can best attack the fire while keeping firefighters safe.

When developing a strategy, a series of questions are addressed, most importantly:

- *What is burning?* The types of fuels involved have a great effect on a fire's behavior. Different vegetation requires different strategies, based on size, thickness, and dryness. A fire burning in sagebrush likely will require a different plan of attack than a fire burning in ponderosa pine trees.
- *What is nearby?* The firefighting strategy for a forest or rangeland fire will differ from that for a fire near structures. Other factors are also important, such as the presence of people, pets, or livestock. This information is used to choose tactics and to allocate resources in order to save lives, homes, personal property, and valuable natural resources.

Below are 10 of the techniques firefighters use to put fires out quickly, completely, and safely.

1. CONTROL LINE

One of the most important components of wildfire suppression, *control lines* are simply the boundaries—natural or manmade—that firefighters employ to control how and where a fire spreads. A rocky ridge or river can serve as a natural control line, or firefighters can establish a manmade one by, for example, clearing out an extended line of brush. Within this overarching definition, a *fire line* is when the barrier is scraped down "to mineral soil," a *scratch line* is a preliminary line built in a hurry, and a *wet line* is when the area has had flame retardant or water applied. And when you hear on the news that a fire is "X percent contained," this is generally what they're talking about—that some percentage of the fire's perimeter has a control/fire line. Sadly, because fires can sometimes jump the barrier, this means even a 100 percent contained fire can start up again.

2. BURNING OUT

When establishing control lines, digging a small ditch and pulling up some plants isn't always enough. To create a sturdy, fuel-free barrier, firefighters may use small torches to burn the brush just inside a control line. A *burnout* is one of several ways to bolster a control line and further prevent a blaze from escaping the established boundaries.

3. BACKBURN

A *backburn* is similar to a burnout, but requires a slightly more sophisticated technique. Once a control line is established, firefighters may set a controlled blaze downwind of the main fire, just on the inside of the control line. Firefighters then push the new blaze back toward the main fire, burning up all the fuel that lies between the fire and the control line.

4. FLANKING

For a wildfire small enough to be snuffed out using a direct attack, firefighters may begin their assault on the blaze from behind. Starting from already burned earth, the firefighters will typically work their way around the edge of the fire to spray the flames as they make their way around the entire perimeter of the blaze.

5. HOT SPOTTING

Hot spotting is the term used to describe the extra attention given to the most active and dangerous portions of a wildfire. The crews fighting the fire size up the parts of the blaze most likely to spread and try to devise the best strategy for keeping these areas in check. Hot spotting may also involve diverting extra manpower to the task of stamping out embers and spot fires that blow or erupt from the hottest part of the fire.

6. KNOCK DOWN

Whereas hot spotting refers to the assessment of a fire's condition, *knocking down* is all about action. The knock down strategy is employed when firefighters decide that a certain hotspot needs to be suppressed immediately. To diminish the section of a fire deemed to have grown too hot, too active, or too large, fighters directly apply some combination of dirt, water, or retardant to that section.

7. COLD TRAILING

While a fire is being attacked from the front or side, other firefighters may be involved in *cold trailing*, the task of combing through already scorched ground in the wake of a moving wildfire. The point is to make sure no hot or glowing embers remain, since leftover coals can be blown around and flame up again.

8. AERIAL ATTACK

If significant exposed water sources are nearby, planes and helicopters can scoop up buckets of water and carry them to be dropped atop the blaze. The water is often mixed with a foam retardant before being dropped [PDF]. The foamed water acts as a more effective barrier to the spread of fire and also insulates fuel that has not yet burned.

9. FIRELINE EXPLOSIVES

When setting control lines or firelines, firefighters may even use explosives to break up dense brush and fallen trees. Explosives can also be used to fell trees whose spread might help a fire jump across a control line. During a large or fast-moving fire, explosives are employed mainly for efficiency purposes, as they can save precious time when firefighters need to contain a fire quickly.

10. MOP-UP

It's called *mop-up* when firefighters go back and clean up along a completed control line. Mop-up consists of dousing any embers and spot fires that have made their way across control lines. It also involves protecting still-vulnerable fuels using a burnout (if they're permanently situated) or by simply moving them.

CHAPTER 8

Support Measures in Forest Fire Control

Introduction

Most forest fires are started through human activity. Avoiding forest fires can be achieved through various means, but in the end a combination of different measures offers the best protection. In the next pages technical preventative measures will be introduced.

Fire fighting reservoirs

Water is still the main way to extinguish forest fires. Therefore in large, contiguous and fire endangered forested areas it is necessary to have, or build and maintain, a fire fighting water supply system within suitable water courses or to create artificial reservoirs for water extraction. When creating new reservoirs an agreement is needed between the forest owners, forest authorities and the fire service. It is important that these extraction points are sufficiently identified and easily accessible by fire engine.

Along with natural watercourses, deep or shallow wells, artificial ponds, dams, underground water tanks (cisterns) or auxiliary water supplies in or near forests can be used as firefighting reservoirs. The functionality and accessibility of these water extraction points must be checked regularly.

Infrastructure

In order for fire engines to reach a forest stand it is important that the roads can bear heavy vehicles. Similarly turning and passing bays and sufficient clearance must be maintained or created. Also as with

Figure : Fire fighting reservoir.

the creation of water reservoirs, this forest infrastructure should be agreed with the forest authorities, owners and fire service.

Machinery and equipment

Nowadays, alongside mobile fire extinguishing equipment, fighting forest fires continues to be hard manual work for as many people as are available to fight the fire. Therefore maintaining appropriate fire-fighting tools and machinery is the responsibility of forest enterprises of all ownership type in areas with a medium to high forest fire risk. These include hand tools such as spades, shovels, fire beaters and axes as well as transport vehicles or tractors and ploughs suitable for working in forests. Such tools and machinery are either kept by the forest enterprise themselves or contracted in.

Forest Fire Monitoring

Since the introduction of automatic, camera supported forest fire observation systems, the number of forest fires has not reduced but their extent has. The early and exact identification of forest fires and a fast notification has ensured that the technical team can be onsite quickly and can quickly begin to fight the fire. An optical sensor system used for the early detection and monitoring of forest fires is described in a separate article.

Figure : Fire watchtower.

Aerial Surveillance

Aerial surveillance flights are a possible means to detect forest fires at an early stage during times of high fire risk. Along with regulatory authority surveillance flights (e.g. fire service air service in Lower Saxony), commercial airlines or private amateur pilots can be used to fly pre-determined routes. This form of collaboration is practiced in Saxony, Bavaria and Brandenburg. In Rhineland-Palatine when the danger level is above level 4, all airports are contacted and pilots are asked to help with early detection. Besides the early detection and location of forest fires, this aerial support can also be helpful in directing the operational forces on the ground.

Mapping

Forest fire fighting maps at a scale of 1:50.000 using the UTM geographic coordinate system are the basis for all those fighting forest fires. All important elements such as fire fighting water points, towns etc are shown on these maps. The depiction follows the tactical symbols of the fire service. These maps are prepared by the authorities and are updated at least every five years. In the case of large fires, district maps at a scale of 1:25.000 or even 1:10.000 may be used. Besides orientating out of town operational teams, these maps serve as guides in the central control offices. When using automated camera monitoring systems, maps and photos of the fire location are digitally produced and given to the ground forces.

Since 2008, the state forest administration, fire service as well as communities and districts in North Rhine-Westphalia have been equipped with the "Forest Fire Emergency Response Map". These maps at a scale of 1:25.000 cover the entire state and show the roads capable of bearing heavy vehicles and turning places in forests, still and running water courses as well as areas with particular dangers including the "Kyrill" storm damaged areas in Rhineland-Palatine the "Forest Rescue Service" provides maps to operational teams which contain information about heavy load bearing capable roads and the location of rescue points, which are also identified on-site. Additional forest fire relevant information is not identified.

Communication equipment

Forest fires can only be quickly and successfully fought with functioning communication between the fire service and forest authority operational teams. Mobile telephones and radios are required. Up-to-date telephone lists are also needed.

Annual checks are carried out before the start of the fire season in Mecklenburg-Western Pomerania to test that forest fire fighting maps, telephone, mobile and address lists as well as response plans are up-to-date.

Emergency and deployment plans, Control of operations

Before the outbreak of a fire, emergency and deployment plans have to be created. In the regions of Mecklenburg-Western Pomerania at risk from forest fires this is the responsibility of the district working group "Forest fire protection".

In addition to updating plans, in the run up to a fire, suitable premises for an operational headquarters can be identified if a command vehicle will not be used as the control centre. Besides technical equipment (computers, printers), these premises should also be equipped with sufficient communication means (emergency services radio, mobile telephones, landlines, receivers). A good road connection and a helicopter landing site is also required. As the press and media representatives have to also be reckoned with during a fire, a separate room should be made available for them to work in if the fire will last for some time.

Cooperation and joint exercises

Figure : Forest fire exercise.

Collaboration between forest owners, the administration and different branches of the fire and emergency services is of special importance when it comes to forest fires. In the forest fire endangered areas of the new German states, representatives from the forest authority, forest owners, fire service, fire and emergency services, operational directors and other professions in the district work together in district "Forest Fire Protection" work groups. These work groups organise joint education and training courses as well as exercises, evaluate forest fire events and develop and update the emergency plans.

Alongside the technical aspects, getting acquainted with each other plays an important role. Sharing common experiences and getting to know one another during the evaluation, planning and implementation exercises help guarantee a collegial relationship exists when required. In this way mistakes are avoided, decisions are made quickly and decisively and the area of burnt land reduced.

Mecklenburg-Western Pomerania	Brandenburg	Schleswig-Holstein
• The district forest officer undertakes certain forest fire protection duties for several forest offices (according to district boundaries and forest fire danger ranking) • District forest fire protection representatives coordinate the collaboration between the forest authority and the district or city in the forest fire protection working groups • Annual readiness, operational and emergency plans (Forest authority plans involving the federal and non-state forest authorities and forest owners) • Annual check of • forest fire fighting maps • telephones, mobile phones, address lists • emergency plans • Forest fire fighting exercises • Annual radio and alarm exercises • In A and B danger classified areas the education and training is every two years, in areas in danger class C, every three years	• · Forest fire protection representatives in the Forestry Office are responsible for: • The annual development of the forest fire emergency plan • Coordinating forest fire prevention • Representing the Forestry Office in district and city committees • Acting as the forestry specialist in command during large scale events or catastrophes • Coordinating public education about forest fires for the Forest Office • Developing a Forest Protection work group in areas with A and A1 danger classes (Members: district fire officer, Forest fire protection representative of the Forestry Office, and representatives from the state forest, federal forest, communal and private forest) • Annual updating of the forest fire emergency plan (responsibility and accessibility, available manpower and resources • Annual forest fire protection exercises	• Creating emergency and operational plans at the district level (considering site specific risks) • Every 10 years estimating risks at a joint meeting of representatives of forest owners, the district and local forest authorities, local regulatory authorities and communities • During periods of high risk developing, teaching and practicing a detailed emergency and operational plan, otherwise collating operational resources and a thorough specification of defence structures • Integration with the Civil Defence's emergency plans

FIRE FIGHTING WATER STORAGE SOLUTIONS

Automatic fire sprinkler systems are installed for two main reasons; property protection or life safety. In both cases their efficiency in controlling and extinguishing fires has protected lives and the environment for over one hundred and twenty years.

In both cases their efficiency in controlling and extinguishing fires has for over one hundred and twenty years protected lives and the environment and it should be accepted that all sprinkler systems will protect lives by restricting fire spread and protection escape routes.

Property protection systems are often installed at the request of the building occupier's insurer to protect the business by ensuring the building and contents are protected against fire.

Some systems are installed to comply with building or fire regulations primarily to protect employees, the public and fire fighters from the risks of fire. This type of system should be more accurately referred to as 'enhanced availability' sprinkler systems but are more commonly called 'life safety' sprinkler systems and include a range of added features which reduce the possibility that the system will ever be non-operational due to scheduled service and maintenance being carried out.

In both of these cases an essential part of the system is the water supply. This can take the form of a direct supply from the local water service main with or without a booster pump or, more reliably, having water stored in a tank with a pump or pumps to deliver the water to the sprinkler system.

The most common arrangement is a single water storage tank with two fire pumps, each capable of meeting the needs of the sprinkler system.

For enhanced reliability, the water storage tank can be split into two half-capacity tanks. This ensures that there is always a water supply available to the sprinkler system, even when one tank (or any of its equipment) is being serviced or maintained.

PROVISION AND MAINTENANCE OF WATER SUPPLIES FOR FIRE FIGHTING—CODE OF PRACTICE

Water supply requirements for firefighting will depend on various factors like type of construction, nature of occupancy, type and quantities of materials handled or stored, process risks involved, neighbouring fire risks, fire detection and alarm arrangements, time likely to be taken for the fire brigade(s) to come into operation and so on. All these factors may differ very widely and so will be the requirement of water supplies. Requirement for each area may therefore vary and has to be worked out jointly by water supply engineers and fire experts after considering these factors.

The essential requirements for any source of water for firefighting are as follows:

Ready availability of the supply at all times,

Sources easily approachable and workable by the normal fire appliances, and

Sources located within a workable distance which should in no case be more than 1 km from the risk to be protected.

Normally such requirements are met with two types of water supply namely piped water supply (hydrant system) and static water tanks. However, on some occasions natural and man-made sources if available could be utilised for fire fighting in emergencies. The commonly found natural sources are rivers, canals, streams, ponds, lakes, pools, spring wells, etc, and man-made sources are reservoirs, pooling water basins etc. Therefore, while planning any system for water supply requirements such sources may also be considered and suitable arrangements made for their effective and quick use by the fire services like all weather approach routes, hard standing platforms, provision of suitable sumps, portable pumps in deep wells etc. In case of overhead reservoirs suitable number of 100 mm round thread or 63 mm instantaneous outlets may be provided on the main, outlets being controlled by separate control valves.

WATER SUPPLY REQUIREMENTS

For Towns/Cities

The fire risk in any town/city is seldom uniform, throughout and it may vary widely in different areas. It may be the lowest in well laid out predominently residential locality, with small shopping centres and go on increasing in thickly populated congested areas, commercial centres, warehouses and industrial complexes. In addition to normal peace time requirements, some vulnerable towns/cities may require very large extra quantities to deal with special situations that may arise in war time or other emergencies to which the towns/cities may be exposed. These are generally termed as 'civil defence' towns. With so may variants, it may be desirable to work out the water supply requirements area-on locality-wise after considering various factors, but, this may not be possible in each and every case.

The overall requirements may, therefore, be worked out on population basis. But, while working out the detailed distribution system, the peculiar requirements for each area shall be examined and extra provision made where required.

It is recommended that generally for towns/cities, water for firefighting shall be provided at the scale of 1800-l/min for every 50 000 population or part thereof for towns up to 3 lacs population and an additional 1800-l/min for every 1 lac population of more than 3 lacs. The requirement should be on the basis of 2 hours duration.

In addition to the population criteria, it should be ensured that sufficient water at the above scale is made available within every 1 km2 area of the city/town and it should be ensured that it is equitably distributed. In the case of smaller towns with population of 1 lac and below the total requirements should be doubled.

For high risk areas like bazars, commercial centres, high rise buildings warehouses, industrial complexes, etc, extra provision shall be made considering each risk area separately. For extra high hazard risk like petrochemical chemical complexes extra provisions, on the external periphery shall be provided on. The provision in respect of highrise building are given in IS 3884 : 1989, there should be at least one1 static tank of 220 000 litres capacity for every 1 km2 area.

For civil defence towns/cities, the scale as prescribed under 4.1.2 may be doubled, both in respect of per minute and total requirements. The extra provision on this account shall be made in the form of static sources as far as possible.

For Industries

Industries vary vary widely in size and fire risk. There are small scale industries located in small shed(s) employing only a few persons. On the other extreme, there may be big industries which are spread over hundreds of acres and employ thousands of persons. Risk wise also, there are low to medium risk industries handling and storing mostly non-combustible or ordinary combustible materials against the very high fire risk industries handling and storing sizeable quantities of highly combustible and flammable substances. Therefore, the guiding factor shall be to determine the number of fire fighting jets which are likely to be employed in case of a major fire outbreak in the industry. For this purpose, the risk category of the industry should be first determined, which will quantify the number of fire fighting jets vis-a-vis the quantity of water required to be provided.

The industries may be divided into three main risk that is light hazard (covering group A to F), moderate hazard (covering group G and G-1) and high hazard (covering group H and J) as follows:

For all risks coming under light hazard category and with single unobstructed floor area not exceeding 1 000 m2, it shall have water storage for firefighting equivalent to 1 hour pumping capacity minimum, when using two strong water jets simultaneously—each jet consuming 600 lpm. Where the area exceeds 1 000 m2 additional quantity at the rate of 50 percent of above shall be provided subject to a minimum of 2 hours pumping capacity.

For all risks coming under moderate hazard category and with single unobstructed floor area not exceeding 1 000 m2, it shall have water storage for firefighting equivalent to 2 hours pumping capacity minimum, when using 4 strong water jets simultaneously—each jet consuming 600 l pm Where the area exceeds 1 000 m2, additional quantity at the rate of 50 percent of above shall be provided, subject to a minimum of 2 hours pumping capacity.

For all risks coming under high hazard category and with single unobstructed floor area not exceeding 1 000 m2, it shall have water storage for firefighting equivalent to 3 hours pumping capacity minimum, when using 6 strong jets simultaneously—each jet consuming 600 lpm. Where the area exceeds 1 000 m2, additional quantity at the rate of 50 percent of above shall be provided subject to a minimum of 4 hours pumping capacity.

However, on single unobstructed floor area not exceeding 1 000 m2, it shall have water storage for firefighting equivalent to 4 hours pumping capacity minimum, when using 8 strong water jets simultaneously—each jet consuming 600 lpm. Where the area exceeds 1 000 m, additional quantity at the rate of 50 percent of above shall be provided subject to a minimum of 4 hours pumping capacity.

The quantity of water requirement as mentioned above, shall be stored in the form of one or more static tanks.

NOTE—In case of one tank it should be in two equal interconnected compartments.

In case of an efficient municipal fire hydrant system, the quantity of water required to be stored in static tanks could be halved, however the capacity in any static tanks shall not be changed.

In case of any industrial unit which has its own fire hydrant system installed, the entire quantity of water could be stored in a single large tank/compartmental tank/reservoir with a well designed ring main of adequate size on which hydrants can be installed.

For flammable liquid tank farms, the minimum water requirement for firefighting shall be calculated separately for use of foam, and for cooling of the affected tanks and also the surrounding tanks, added together.

For tank farms with fixed foam installation system, the capacity of water tank shall be as follows:

At the minimum rate of water solution calculated at 4.5 lm2 of liquid surface area (tank surface), for a minimum time of 30 minutes.

If fixed foam monitors are installed, an additional quantity of 900 litres per minute for at least another 30 minutes duration shall be provided, if number of monitors are many, provision for at least two monitors at the rate of 900 lpm each to be calculated.

For cooling purpose of adjoining tanks, additional quantity of water shall be provided at the rate of 10 lpm/m2 of exposed surface of the next tank, for a period of at least 1 hour.

Process Units

Water requirement for process units shall be depending upon the risk.

Administrative buildings in factories etc shall be considered as light hazard, and the water quantity and number of hydrants catered accordingly.

Storage Risks

For storage risks engineering stores shall be classified as light hazard category. General stores as ordinary hazard and combustible or flammable stores as high or extra high hazard category. In case of such combined risk, the risk category will be decided by its predominent risk.

STATIC WATER TANKS

Static water tanks are very useful sources of water supplies for major fire fighting. They can be provided to supplement the first hydrant system or independently where provision of fire hydrants is not feasible economically or due to various other reasons. Static water system has many advantages over the hydrant system. This system is much more economical, more reliable, less prone to damage or defects and helps to build huge reserve to meet any eventualities. While planning any water supply system for fire fighting in towns, cities and industries, suitable provision of static water tanks shall always be kept in view. For civil defence towns/cities and all type of industrial establishment, at least 50 percent of the total water supply requirements for firefighting shall always be in the form of static water tanks.

Static water tanks for firefighting should be underground with water at ground level. They can be of any shape and dimensions, but the depth shall normally not exceed 2·5 m. For multi-storey buildings terrace tank may be provided.

NOTE—In case of deeper tanks, submersible pumps be provided.

For towns/cities and other locations where they are accessible to public, the static tank should be completely covered and provided with suitable manholes for lowering of the suction hoses.

Inside industrial establishments the static water tanks can be of open type having small parapet walls of height not exceeding 50 cm above ground level. Such tanks may, as far as possible, conform to Fig. 1. Where extra protection is required, barbed wire fencing, etc, may be provided with suitable openings, for lowering of the suction hose.

All static water tanks for firefighting shall normally be located within 100 m of the risk to be protected.

The static water tanks shall be easily approachable by all types of the appliances held at the fire station(s) providing fire cover in the area. All weather approach road of adequate size shall be provided. Provision of suitable number of manholes shall be made available for inspection, repairs and inspection of static tanks, etc.

Cement concrete platforms (hard standings) shall be provided at suitable locations around the static water tanks for prolonged operations by the fire pumps.

Each static water tank should be provided with sump(s) to allow the use of the total quantity of water for fire lighting and also to facilitate maintenance and repair of tanks. The sump(s) may be of the size 1 m × 1 m × 45 cm (depth). The sumps should be located on the side(s) from where the fire pumps are to come into operation.

U-shaped steel bar steps or any other suitable arrangement shall be provided for men to enter the static water tanks as and when required.

All static water tanks shall be provided with suitable filling arrangements to make up the evaporation losses, refilling after cleaning/ repairs and for replenishing water supplies during fire fighting operations. The filling arrangements shall be of the maximum capacity possible and the tanks shall be connected by the biggest size of mains available in their vicinity at the rate of not less than 1 000 l/minute.

Suitable indicating plates shall be fixed to the nearest wall or a suitable posts erected for the purpose near each static water tank. The indicating plates can be made of vitreous enamelled mild steel, cast iron, aluminium alloy or plastic. The plates shall be yellow in colour with letter 'SWT' and the capacity in litres marked in black.

INSPECTIONS, CHECKS AND MAINTENANCE

Periodical checking and proper maintenance of water supplies is as important as their provision. Since special water supply systems for fire fighting are not in daily use, many times, defects developing over a certain period may not be observed unless periodical checks are exercised. The system failing to come into operation at the time of any fire can defeat the very purpose for which the system has been provided and also result in very embarrassing situations to the fire and other administrative authorities of the area. Such situations are to be avoided at all cost and this can be ensured only when all parts of systems are constantly checked and maintained in perfect working order all the time.

Reservoirs

Required level of water shall always be maintained and this shall be checked daily by the pump operators and incharge water works in each shift before assuming charge.

A monthly inspection of the reservoirs shall be carried out by a responsible officer of the water works to ensure that the reservoirs are constructionally sound and there is no seepage or leakage from any portion. At this time, note may be taken of any accumulation of mud, sand, weeds and other undergrowth which can reduce the capacity of the reservoir and obstruct free flow of water. The required remedial action may be taken.

All reservoirs shall be periodically cleaned, periodicity for such cleaning to be determined and laid considering the local conditions. Before emptying any reservoir for cleaning and/or repairs, the other compartment/reservoirs shall be kept full and duly connected to the system.

Static Water Tanks and Natural Sources

Static water tanks and other natural sources for fire fighting shall be inspected as often as possible. The periodicity for such inspections shall be determined and laid down depending on the local conditions. However, for cities and towns, each static or natural source shall be inspected at intervals not exceeding one month and for industrial establishments once every week. The record of all such inspections shall be maintained and remedial action for any defects observed taken immediately or priority basis.

The following action shall be taken during the monthly/weekly inspection:

Level of the water supply shall be checked and noted.

Causes for any unusual drop in the level may be investigated and reported to higher authorities for further investigations and remedial action.

In the case of static water tanks drop of level due to evaporation and other normal causes, it shall be made good by opening the filling valves. In the case of open static water tanks, sonic water shall be allowed to overflow to avoid breeding of mosquitos, etc.

Static tanks, sumps and jack wells shall be examined for any signs of cracks or other constructional damages. Accumulation of mud, sand, weeds and other undergrowth shall be noted and action taken for their removal.

Condition of approach roads, hard standing platform, parapet walls, fencing, manhole covers, indicating plates, etc, shall be examined and recorded.

Any obstruction in the approach of the fire appliance to the sources of supply shall be removed.

All static water tanks, sumps and jack well shall be emptied, cleaned, repaired, and whitewashed at periodical intervals. Open static water tanks shall be so cleaned once every 6 months For the covered static water tanks, periodicity may be determined according to the local conditions, but they shall be thoroughly cleaned at least once every 2 years.

While emptying any static water tank for cleaning or repairs, it shall be ensured that the other neighbouring tanks are full and in serviceable condition. As far as possible, only one tank shall be taken for repairs and maintenance at a time.

The use of static water tanks and other natural sources shall be made in drills and practices as often as possible. Any difficulties observed in approach and working of the fire appliances recorded and remedial action taken.

Portable water tank

A **portable water tank** is a temporary collapsible tank designed for the reserve storage of water in firefighting, emergency relief, and military applications. These tanks can be either supported or unsupported.

Figure : Fire Department using a Portable Water Tank in South San Jose

The supported tanks have a steel or aluminum frame and range in size from 600 to 5000 US gallons or larger by custom design. Portable water tanks are also unsupported such as self-supporting tanks (onion tanks), blivets and pillow or bladder tanks and are available in sizes ranging from 100 US gallons (380 L) up to 80,000 US gallons (300,000 L).

Usage/Deployment

It is primarily used in rural areas where fire hydrants are not available. They are carried on water tenders and are deployed at the scene of a fire during a shuttle operation. A portable water tank is usually set up near or front of an attack engine, or possibly next to a supply engine. This enables tenders to quickly drop off their load of water and return to the fill site as soon as possible. They are designed to be set up in around a half-minute with two firefighters. The engine may then use suction hose to draft the water in the tank.

Wildfire Technology: Tackling the Spread of Wilderness Fires with Digital Tech

Wildfires are among the most destructive forces on Earth. When these fires burn into towns and cities on the edges of forests or wildlands, the economic damage and potential loss of life can be massive. Meanwhile, hotter global temperatures and extreme heat waves increase the danger from wildfire spread.

The technology used to tackle large area fires has become more sophisticated to respond to this threat. Firefighters form highly-trained ground and air teams. Helicopters and air tankers carry water, suppressant foam, and flame retarding chemicals direct to fire hotspots.

The State-Of-The-Art-Tech Behind Fighting Fires

How are today's firefighters tackling massive wildfires? They're doing it by using the most sophisticated technology. Whether it's with a modified jumbo jet or state of the art thermal imaging cameras, we take a look at the tech behind fighting some of the worlds most devastating fires.

Teams tackling the California wildfires used the Global SuperTanker – a modified jumbo jet that carries almost 73,000 liters (19,200 gallons) of fire retardant – alongside the S-64 Aircrane heavy-lifting helicopter, which carries 10,000 liters (2,200 gallons) of water.

These air tankers are used to put out flames and monitor fire spread through sensors and video feeds coupled to GPS data. Fed into sophisticated computer modelling software, they can help predict the fire's behavior and possible spread patterns. Computer models are able to provide small area coverage only meters wide, mapping terrain and airflow. Fires are especially responsive to wind conditions and these tools allow firefighters to determine the passage of flames at ground level.

But manned aircrafts need a large investment in maintenance and crew training. Command and control centers are turning to unmanned aerial vehicles (UAVs) to keep costs down and provide additional capabilities. Small quadcopters and larger fixed-wing aircraft can fly over fires for much longer periods.

Smoke can obscure the ground for days at a time or pose a severe inhalation risk to air crews – However, this is not a problem for drones. Onboard high definition, infrared and thermal imaging cameras can provide direction to ground teams, spot vital infrastructure (including power or water lines), and identify dangerous or flammable objects.

Infrared and thermal cameras can see through smoke to monitor ground teams and let them know when conditions change. UAVs can provide georeferenced aerial images, heat maps, and temperature scales of fire zones. Specialist drones can even carry hoses to less accessible areas. In the future, swarms of autonomous drones could be used to track wildfires and spot fire spread.

Thermal imaging technology has become widespread and less expensive to use. Handheld cameras and devices that attach to smartphones allow firefighters to see through smoke and find active fire hotspots, or undergrowth that is burning without producing smoke.

Identifying these hotspots allows crews to target the most active and dangerous parts of a willdfire and divert manpower to tackle it more effectively.

From the Amazon basin to northern Siberian forests, the problem is spreading. Hotter dryer conditions mean wildland fires are more likely to spread in regularity and scale. Professor Johann Goldammer, a leading authority on wildfires and director of the Global Fire Monitoring Centre (GFMC) at Germany's Max Planck Institute of Chemistry, says climate change is creating longer, more severe fire seasons. "A region like the West Coast of the US is experiencing an increasingly longer fire season that now seems to be all year long," said Prof Goldammer.

Human behaviour is adding to that risk, he added. "We are modifying the land by clear cutting, inhabiting, entering, industry, agriculture and so on. With this, the fire regimes and the vulnerabilities of ecosystems are changing."

From January 1 to July 30, 2019 there were 25,619 wildfires in the US, according to the National Interagency Fire Center. Some 3.2 million acres were burned in that period. In California alone, there are more than two million properties at high to extreme risk from fire damage. In 2018 the Camp Fire in northern California became the deadliest and most destructive fire on record in the state. At least 88 people died and 18,800 structures were destroyed. The economic losses were estimated at between $8.5 billion and $10.5 billion.

For the crews tackling these blazes the risks are high too. Air crews are regularly prone to incidents and 24 per cent of firefighter deaths betwen 2006 and 2016 were due to plane and helicopter crashes. This highlights the wisdom of expanding drone fleets to provide support to fire teams.

Other technologies that can save lives include surival items like fire blankets. Often the last line of defense for trapped firefighters, a Sunseeker fire blanket can withstand temperatures of 3,000 degrees fahrenheit. First response firefighters who parachute into wildland fires – known as smokejumpers – are among those who need them most.

Their training at the US Forest Service now includes the PARASIM virtual parachute jump simulator to practice for real-life situations. PARASIM provides the same performance as a parachute – where trainers can change wind speed, weather conditions and terrain types – inside a realistic 3D simulation.

Augmented reality helmets that include breathing apparatus alongside computer vision-aided displays are adding to the ground team's capabilities. Thermal cameras inside the helmets mean that firefighters can operate in environments completely obscured by smoke. At the same time they can wirelessly transmit information on what is happening at the frontline to command and control points.

Robots are also making a difference. The Smokebot was developed by a Swedish university to assist fire and rescue services. It collects data in environments with reduced visibility using radar, a laser scanner, a thermal camera and gas sensors. Smokebot can help in forest fire situations mapping large areas filled with dust or smoke, where it is too risky to send in rescue personnel.

Finally to get the bigger picture on fire tracking and monitoring, the US Forest Service and US Geological Survey uses data from the Landsat Earth-observing satellites. Data gathered from every major fire in the country since 1984 has been fed into computer models to help predict and prevent wildfires.

Forest thinning and controlled burns are used to eliminate fuel sources before they can ignite. "We use the satellites to inform decisions on where to stage assets across the country," said Brad Quayle of the Forest Service's Geospatial Technology and Applications Center. "When there's high competition for firefighters, tankers and aircraft, decisions have to be made on how to distribute those assets."

Earth-observing satellites commonly detect wildfires in wilderness areas. Their cameras and remote sensors are used to estimate the fire's evolution and provide situational awareness that saves lives.

The Visible Infrared Imaging Radiometer Suite (VIIRS) onboard the Suomi NPP satellite gives near real-time data to NASA's Fire Information for Resource Management System (FIRMS) on active fires across the world.

Fire Fighting Equipment

Firefighting equipment is generally comprised of hundreds of different pieces of equipment and technology.

A categorized list of such equipment might look like this:

- Extinguishers – using water, dry chemical powder, foam, carbon dioxide, or other substances
- Fire fighter's gear – fire resistant outer suit, inner lining suit, inner and outer gloves, woolen socks, fire resistant shoes, helmet (with or without headlight), communication system, and breathing apparatus
- Installed fire fighting system – main water supply network, hydrant, sprinkler (automatic or manual), emergency fuel and electric cut off system, portable generators, and portable pumps
- Communication equipment – walkie-talkie, radio, broadcasts, landline telephone, sound powered telephone, cell or mobile phone, and voice pipe
- Fire detection and alarm systems – smoke, frame, spark detectors, local, centralized automatic, semi-automatic, and manual fire alarm systems
- Other accessories – water and sand bucket, shovel, hammer, fire axe, cutters, hooks, fire blanket, emergency lifesaving apparatus, and emergency lights

Firefighting equipment is equipment designed to extinguish fires or protect the user from fire. It may be used by trained fire fighters, untrained users at the scene of a fire, or built into a building's infrastructure (such as a sprinkler system).

Firefighting equipment includes not only fire hoses and fire extinguishers but also fire-resistant protective clothing, fire-resistant gloves, respirators, and communication equipment

Smartphones in professional operations

Smart fire services are looking for ways to include smart device users in operations – in a secure, controlled way. One way is through a professional group communication application such as Tactilon Agnet.

The Tactilon Agnet push-to-talk app can connect teams seamlessly whether the members are carrying a TETRA/Tetrapol radio or a smartphone – or even an analogue radio or a smartphone.

New technologies such as Artificial Intelligence AI, Virtual Reality VR, and Augmented Reality AR offer new options for public safety apps, such as

- Tracking of vehicles, assets and resources

- Messaging in talk groups
- Camera and video recording, streaming and sharing
- Call taking
- Command and Control in the office and in the field.

SAFEcommand - a better way

A better way would be to automatically get the vital information needed for tackling an incident, without having to shuffle manually through stacks of paperwork that may be out of date. This is exactly what SAFEcommand can deliver.

SAFEcommand is a system that provides real-time location intelligence for staff and vehicles.

Choose hybrid communications

Tactilon Dabat is a smartphone and a full TETRA radio in one device. It is smart, strong and secure.

- Use it as a TETRA radio for critical communications
- Use it as a smartphone for mobile applications and special professional apps.

In addition, Tactilon Dabat offers Hybrid Roaming. Running Tactilon Agnet app, Tactilon Dabat can automatically move from using the TETRA network to using a 4G/5G network or vice versa.

Wildfire suppression

Wildfire suppression is a range of firefighting tactics used to suppress wildfires. Firefighting efforts in wild land areas require different techniques, equipment, and training from the more familiar structure fire fighting found in populated areas. Working in conjunction with specially designed aerial firefighting aircraft, these wildfire-trained crews suppress flames, construct fire lines, and extinguish flames and areas of heat to protect resources and natural wilderness. Wildfire suppression also addresses the issues of the wildland–urban interface, where populated areas border with wild land areas.

In the United States and other countries, aggressive wildfire suppression aimed at minimizing fire has contributed to accumulation of fuel loads, increasing the risk of large, catastrophic fires.

Australia

Wildland fire, known in Australia as bush fire, has played a major role in Australia due to arid conditions. Notable fire services tasked with wildfire suppression include NPWS (National Parks and Wildlife Service, NSW), the New South Wales Rural Fire Service, the South Australian Country Fire Service, and the Victorian Department of Environment Land Water and Planning (DELWP) and the Country Fire Authority.

Canada

Canada contains approximately 3,964,000 km^2 of forest land. Seventy-five percent of this is boreal forest, made up primarily of coniferous trees. More than 90 percent of Canadian forest land is publicly owned, and the provincial and territorial governments are responsible for fire-suppression activities. The Federal Canadian Interagency Forest Fire Centre (CIFFC) provides operational fire-control services and links to all provincial and territorial fire agencies.

During a typical year there are over 9,000 forest fires in Canada, burning an average of 2.5 million hectares (ha) or 9,700 square miles (25,000 km^2). The number of fires and area burned can vary dramatically from year to year. Average suppression costs are $500 million to $1 billion annually.

In Canada, two-thirds of all forest fires are caused by people, while lightning causes the remaining third. Despite this, lightning fires account for over 85 percent of the area burned in Canada, largely because many of the lightning-caused fires occur in remote, inaccessible areas. Currently about ninety percent of forest fires are fought. Generally fires near communities, industrial infrastructure, and forests with high commercial and recreation value are given high priority for suppression efforts. In remote areas and wilderness parks, fires may be left to burn as part of the natural ecological cycle.

United States

Wildfire suppression in the United States has had a long and varied history. For most of the 20th century, any form of wildland fire, whether it was naturally caused or otherwise, was quickly suppressed for fear of uncontrollable and destructive conflagrations such as the Peshtigo Fire in 1871 and the Great Fire of 1910. In the 1960s, policies governing wildfire suppression changed due to ecological studies that recognized fire as a natural process necessary for new growth. Today, policies advocating complete fire suppression have been exchanged for those who encourage wildland fire use, or the allowing of fire to act as a tool, such as the case with controlled burns.

Objectives and Risks

Safety

Figure : Lava flow on the coastal plain of Kîlauea, on the island of Hawaii, generated this wildfire.

Protection of human life is first priority for firefighters. Since 1995, when arriving on a scene, a fire crew will establish safety zones and escape routes, verify communication is in place, and designate lookouts (known in the U.S. by the acronym *LCES*, for lookouts, communications, escape routes, safety zones). This allows the firefighters to engage a fire with options for a retreat should their current situation become unsafe. Although other safety zones should be designated, areas already burned generally provide a safe refuge from fire provided they have cooled sufficiently, are accessible, and have burned enough fuels so as to not reignite. Briefings may be done to inform new fire resources of hazards and other pertinent information.

A great emphasis is placed on safety and preventing *entrapment*, a situation where escape from the fire is impossible. Prevention of this situation is reinforced with two training protocols, *Ten Standard Firefighting Orders* and *Eighteen Situations That Shout Watch Out*, which warn firefighters of potentially dangerous situations, developed in the aftermath of the Mann Gulch fire. As a last resort, many wildland firefighters carry a fire shelter. In this inescapable situation, the shelter will provide limited protection from radiant and convective heat, as well as superheated air. Entrapment within a fire shelter is called a *burnover*. In Australia, firefighters rarely carry fireshelters (commonly referred to as "Shake 'N' Bake" shelters); rather, training is given to locate natural shelters or use hand tools to create protection; or, in the instance of 'burnover' in a tanker or other fire appliance, 'fire overrun' training is used.

Hazards beyond the fire are posed as well. A very small sample of these include: unstable/hazardous trees, animals, electrical cables, unexploded ordnance, hazardous materials, rolling and falling debris, and lightning.

Personal safety is also vital to wildland firefighting. The proper use of PPE (personal protective equipment) and firefighting equipment will help minimise accidents. At the very minimum, wildland firefighters should have proper fire-retardant clothing (such as Nomex), protective headgear, wildland firefighting-specific boots, gloves, water for hydration, fire shelters, eye protection, and some form of communication (most commonly a radio).

Resource protection

Other resources are ranked according to importance and/or value. These include but are not limited to human health and safety, construction cost, ecological impacts, social and legal consequences and the costs of protection. *Defendability* is also considered, as more effort will need to be expended on saving a house with a wooden-shake roof than one with a tile roof, for example.

Ecosystem Changes

While wildfire suppression serves human safety and resource protection, the lack of natural fires can be the cause of ecosystem changes. Notably, across the global grassland and savanna ecosystems, fire suppression is frequently found to be a driver of woody encroachment.

Fires at the wildland-urban interface

"There are three categories of interface fire: The classic wildland/urban interface exists where well-defined urban and suburban development presses up against open expanses of wildland areas; the mixed wildland/urban interface is characterized by isolated homes, subdivisions and small communities situated predominantly in wildland settings; and the occluded wildland/urban interface exists where islands of wildland vegetation occur inside a largely urbanized area."

Figure : Water Tender ANF 10, from the USDA U.S. Forest Service, Angeles National Forest in the San Gabriel Mountains of the Southern California, Los Angeles Area. This Type II Tender is staffed by two crew members.

Expansive urbanization and other human activity in areas adjacent to wildlands is a primary reason for the catastrophic structural losses experienced in wildfires. Continued development of wildland-urban interface firefighting measures and the rebuilding of structures destroyed by fires has been met with criticism. Communities such as Sydney and Melbourne in Australia have been built within highly flammable forest fuels. The city of Cape Town, South Africa lies on the fringe of the Table Mountain National Park. In the western United States from the 1990s to 2007, over 8.5 million new homes were constructed on the wildland-urban interface.

Fuel buildup can result in costly, devastating fires as more new houses and ranches are built adjacent to wilderness areas. However, the population growth in these fringe areas discourages the use of current fuel management techniques. Smoke from fires is an irritant and a pollutant. Attempts to thin out the fuel load may be met with opposition due to the desirability of forested areas. Wildland goals may be further resisted because of endangered species protections and habitat preservation. The ecological benefit of fire is often overridden by the economic benefits of protecting structures and lives. Additionally, federal policies that cover wildland areas usually differ from local and state policies that govern urban lands.

Wildfire suppression equipment and personnel

Wildfire suppression equipment and personnel is part of the science of fire fighting focusing on the use of specialized equipment, training and tactics to effectively control, surround and eventually extinguish a natural cover fire. There are several specially designed tools that through their function and user training, perform specialized tasks that are specific to natural cover firefighting. This is used together in conjunction with the general understanding of the behavior of fire to form a viable plan of attack.

Personnel

PPE or personal protective equipment is generally standardized for a certain type of team or crew. Smokejumpers for example require more protective equipment because of their method of delivery into the fire. Ordinarily, all firefighters regardless of assignment, require durable fire recommended eight inch (203 mm) minimum boots, gloves, Nomex pants and shirt, a hard hat (sometimes full brim), Walkie-talkie, potable water, eye protection (goggles) and a fire shelter.

In the U.S., a firefighter's credentials and level of training is shown on their red card. For example, a sawyer will have a feller rating on their red card designating the capacity size tree they have been trained to fell. A firefighter trained as an EMT may have this certification on their card. Classes under the National Wildfire Coordinating Group (NWCG) and management qualification standards under the Interagency Fire Program Management (IFPM) are generally shown as well. At a minimum, the passing of the pack test is displayed showing the individual is physically able to perform appropriate duties.

There are currently 2 wildland firefighter ratings. Firefighter Type 2, which is the basic firefighter qualification, and is required for most operational positions. The Firefighter Type 2 requires several introductory classes as well as a practical day in the field. The dispatch code for a Firefighter Type 2 is FFT2. The second rating is Firefighter Type 1. The Firefighter 1 rating requires advanced courses in wildland firefighting as well as acceptable performance as a FFT2. Firefighter Type 1 is considered to be a senior and seasoned wildland Firefighter. The dispatch code for Firefighter Type 1 is FFT1.

Hand crews

Typically, wildland firefighting organizations will use large handcrews of 20 or more people who travel in vehicles to the fire incident. Although these crews can vary above or below 20 firefighters, they are generally called twenty-man crews. The designations of these crews in the U.S., defined in large portion by training, are as follows:

Crews
Type I Interagency Hotshot Crew
Type II Initial Attack (IA) Crew
Type II Crew
Type III Crew (none currently in service)

Some personnel are organized into fast attack teams typically consisting of five to eight personnel. Similar to the larger crews, they travel by vehicle.

Smokejumpers

Smokejumpers are highly skilled firefighters specially trained in wildfire suppression tactics. They parachute into remote areas from aircraft to combat wildfires and are equipped to work in remote areas for extended periods of time with little logistical support.

Helitack

Figure : Water drop from helicopter, NPS photo by Harlan Kredit.

The use of helicopter-delivered fire resources varies by agency. Often, helitack crews perform duties similar to other initial attack crews. Two or three firefighters will be dispatched to a newly reported fire. Helitack crews are usually used for initial attack on fires that are difficult for other firefighters to access, or on extended fires that require aerial support in the form of water drops, cargo delivery, crew shuttling, or reconnaissance. A typical initial attack response by a helitack crew involves flying to the fire via helicopter and spending one to three days (although sometimes much longer) putting the fire out before hiking to the nearest road for pickup.

Rappellers

A highly effective way to fight wilderness fire when no roads are nearby is to have wildland firefighters rappel from a helicopter. These firefighters then take suppressive action on the fire or clear a safe landing zone to receive additional firefighters if the fire is too large. Rappellers usually carry 30 pounds of personal gear plus up to 300 pounds of fire gear which is lowered down to them from their helicopter. Rappelling heights can range from 30 feet (in tall, continuous brush) to 250 feet (in timber). When suppression is complete on rappel fires, ground transport is typically arranged to pick up the firefighters at the nearest road. These crews carry chainsaws, hand tools, radios, and can even have 75-US-gallon (280 L) water bags, known as blivets, flown in to help fight the fire. When not rappelling, the crew works as a helitack crew and can fly or hike to any regular fire.

Vehicles

Engines

- Wildland fire engine

Crew transport

- Buggy or crummy
- Crew bus

Tenders

When water is required to refill a fire engine, water delivery is vital. The typical water tender carries 1,200 US gallons (4,500 L) of water to support fire engines. In addition to supplying fire engines directly, tenders may fill water reservoirs for bucket-dropping helicopters when a lake or reservoir is not nearby.

Heavy equipment

Heavy equipment's primary function of wildfire suppression is through the application of heavy construction style equipment to move

Figure : Masticator, which chews through brush and small trees, on the Zaca Fire

large amounts or earth or remove vegetation. Fuel breaks, safety zones, firelines and access to areas that maybe previously were inaccessible may be made. Bulldozers and tractor plows are examples.

Air suppression

In addition to aircraft being used for deploying ground personnel, firefighting groups may utilize helicopters and fixed-wing airplanes specially equipped for use in aerial firefighting to douse areas that are inaccessible to vehicles with water and/or flame retardant chemicals.

Fixed-wing airplanes

- Airtankers
- Lead aircraft
- Smokejumper transport aircraft
- Airlift aircraft

Rotary-wing aircraft

- Helicopters (helitack & rappel)
- Helicopter (water delivery)

Hand tools

A number of tools are used in wildland firefighting. Some examples include:

Figure : Holding a Pulaski, a wildland firefighter from the Angeles National Forest responds to a fire in Altadena, California on July 9, 2006.

- Driptorch
- Fusee
- McLeod (tool)
- Pulaski (tool)
- Fire flapper (tool)
- Quick Release Firehose Clamp (tool)
- Hazel hoe
- Chainsaw
- Shovel
- Very pistol
- Hand held thermal detectors
- Sling psychrometer
- Portable bladder bag
- Two-way radio

Machinery and equipment

There are two types of fire-fighting equipment: portable (eg fire extinguishers, fire blankets) and fixed systems (eg fire hydrants, sprinkler systems).

Portable fire-fighting equipment

The function of a portable fire extinguisher is to deal with small fires that are detected soon after ignition. Fire-fighting equipment should be considered as a means of both prevention and protection. For example, portable fire-fighting equipment can prevent a small fire growing out of control and spreading beyond the area of origin, affecting the means of escape and posing a risk to relevant persons.

The provision of portable fire extinguishers should be as a result of a risk assessment and the requirements of BS 5306-8:2012 Fire Extinguishing Installations and Equipment on Premises. Selection and Positioning of Portable Fire Extinguishers. Code of Practice, which assumes that a fire risk assessment has been undertaken.

In simple premises, having one or two portable extinguishers of the appropriate type, readily available for use, may be all that is necessary. In more complex premises, a number of portable extinguishers may be required.

Basic Fire Fighting Tools Used by Wildland Firefighters

Here is a list of the basic tools, instruments, and equipment issued to active firefighters and necessary to manage either a fire that is prescribed by a forest plan or a wildfire that is under suppression. Having each firefighter equipped with the appropriate hand tool and safety equipment along with a communication link and items for personal comfort under extremely hot conditions are very important.

Wildland Firefighter Hand Tools

Hand tools used by wildland firefighters are always determined by the assignment of that particular person. The numbers and kinds of hand tools used also depend on whether a fire is controlled or out of control and the initial or expected size. I only include the rake and flap, which are necessary under nearly all fire conditions.

A sturdy rake with large triangular cutting teeth is my favorite and called a council fire rake. This tool is designed for fire-line digging. The cutting heads are on a 12"- wide hoe-type frame. It generally consists of four mowing machine cutter blades riveted to a steel frame.

Another popular rake style is called the McLeod fire tool and is another rake-and-hoe combination fire-line digging tool which is popular on mountainous and rocky terrain.

The fire flapper or swatter is always very handy where there are spotting fires near brush and limited available water. They can be a bit heavy but are sturdy enough to do the job of beating and smothering out fires caused by airborne embers floating across the fire line.

The Backfire Torch and Backpack Pump

A backfire torch or drip torch is an essential piece of equipment used to control "fire with fire" when a forest management plan suggests a prescribed burn. This "torch" actually drips a mixture of gas and diesel fuel onto a wick and creates fire on the inner-side of the protective firebreak and the planned burn area. It can also change the direction of an uncontrolled wildfire if used properly.

This initial "dripped" fire is used inside a containing firebreak to manage the fire's rate of spread and to widen the burned "black" area along the area next to the fire break. It does exactly the same thing on a wildfire and is necessary equipment to a wildland firefighter trying to control a fire.

A 5-gallon backpack water pump is a nice bit of additional protection from spotting embers that cross the break and from burning snags and stumps near the fire line. However, it is very heavy, has to be frequently refilled and should only be used by a fit firefighter. This type of pump is best used, along with larger volume capacity pump sprayers, when you have ATV support along the fire breaks.

Wearable Protection for Firefighters

Wearing protective gear is a requirement of most US and state fire protection agencies. Here are the three most important items and should be considered standard equipment on all controlled burns as well as wildfires.

- Wildland fire shirts and pants - the shirt material should be of Nomex quality that has significant heat and flame resistance.
- Full brim hard hat - the hat should have a shell constructed from high-density polyethylene.
- Wildland firefighting gloves - These gloves should have an extra sleeve length made of a fire retardant material.

Fire Shelters for Wildland Firefighters

Wildland firefighting is hard work and done in a high-risk environment. The United States Forest Service requires all of their

firefighting employees and contractors to wear a protective tent called a fire shelter. Both firefighter and non-firefighter can become fatalities during an uncontrolled wildfire in just a few seconds and these "shelters" are not always effective when deployed incorrectly or near heavy fuels.

The fire shelter was developed to become that last piece of equipment you choose to use when conditions and time make survival impossible during a wildfire. The United States still makes shelters mandatory for crews - Canada has discouraged fire shelters.

The new-generation M-2002 fire shelter provides increased protection from radiant and convective heat in wildland firefighter entrapment situations.

CHAPTER 9

THE ROLE OF INDIGENOUS USE OF FIRE IN FOREST MANAGEMENT

Introduction

With devastating wildfires happening more regularly due to climate change, many Native people in U.S. states like California are pointing out that traditional Native methods for managing fire are often suppressed.

Non-Indigenous governments all over the world have stamped out traditional ways of working with fire. But some people are working to bring them back.

Many Indigenous societies use fires in ways that connect ecological, social, and spiritual purposes. For example, the Mebêngôkre of Mato Grosso, Brazil, use fire in their hunt for tortoises. The event is part of an annual festival associated with courtship, youth initiation, and the education of younger generations. The Pemón in Venezuela organize large cooperative groups for tree cutting and burning. This strengthens social ties and passes along knowledge while accomplishing a difficult task.

"Therefore, savanna and forest ecosystems are being protected within Indigenous lands not because they are being 'managed' in a direct and active way, but as the indirect outcome of a healthy social-ecological system," the authors write.

This kind of relationship with ecosystems is a boon to forests and the climate. Mistry, Bibao, and Berardi write that areas controlled by Indigenous peoples tend to suffer less from deforestation and other ecological harms than other comparable land. That means they emit less greenhouse gases. For example, in the Bolivian Amazon, only 0.5 percent of Indigenous territories were deforested between 2000 and 2010, compared with 3.2 percent in the region overall.

However, Indigenous ways of working with fire have come under attack. In California, for example, the Indigenous use of fire was banned outright—even as many Native people in the U.S. were enlisted as firefighters. In other cases, knowledge about these techniques is being lost. On the savannas of Guyana, some Wapishana and Makushi have adopted anti-fire ideas from state resource management agencies and environmental groups. Meanwhile, private cattle owners have increased burns to clear more grazing land, without necessarily employing traditional knowledge about the use of fire. The result is less patch-burning during the wet season—which can reduce the likelihood of large, uncontrolled fries—and more potentially damaging dry-season burning.

1. Community owned solutions for fire management in tropical ecosystems

Environmental management and governance across the developing world is facing an unpredictable and dynamic future, with challenges from entrenched poverty and inequality, slow progress and unintended consequences of national and international development policy, and rapid changes in the natural environment itself. At the same time, there is growing evidence for the fundamental role of Indigenous land-use practices in controlling deforestation, reducing CO_2 emissions and enhancing biodiversity. The United Nations (UN) recognizes that there are over 370 million Indigenous people in at least 90 countries, still undertaking unique practices distinct from those of surrounding dominant societies. Emerging research shows the fundamental role of Indigenous land-use practices for controlling deforestation and reducing CO_2 emissions—analysis of satellite imagery suggests that Indigenous lands have reduced rates of deforestation and habitat conversion, and lower greenhouse gas (GHG) emissions, compared with surrounding areas. Stevens *et al.* , for example, found that in Bolivia, from 2000 to 2010, only 0.5% of the land area of Indigenous territories were deforested, compared with 3.2% overall in the Bolivian Amazon. Rates of deforestation were thus six times lower in Indigenous lands compared with other forests. In another study in the Brazilian Amazon, Petersen & Stevens found that from 2000 to 2012, forest loss was only 0.6% inside Indigenous lands compared with 7.0% outside—thus, more than

10 times lower. Carranza *et al.* show that in the Brazilian *cerrado*, Indigenous lands experienced lower habitat conversion during 2002–2009 than did matched unprotected sites. Similarly, Flantua *et al.* observed that in the western Indigenous Pemón-inhabited sector of Canaima National Park, Venezuela, deforestation rates between 1986 and 2006 were considerably lower, at "0.037% (associated with traditional farming), compared with the 10 km unprotected buffer zone ("0.17%) and average national rates ("0.63%).

The growing empirical data providing evidence for the sustainability of Indigenous peoples' practices in environmental management have prompted much work on power-sharing arrangements, decentralized resource governance and community-based natural resource management , as well as calls to recognize divergent values, participation in political decision-making and equitable distribution of benefits . Yet although there is increased presence of Indigenous peoples and other marginalized local communities in policy mechanisms , Indigenous practice and knowledge is still marginalized in research, development and policy-making circles, primarily as a result of ontological differences, i.e. a mismatch in cultural understandings between Indigenous and non-Indigenous perspectives .

For example, analysis of information sharing in the current REDD+ process in Indonesia found a disconnect between governmental, transnational and domestic civil society organizations . This suggests that multiple perspectives are unlikely to be fully integrated, and that groups may regard one another as information sources lacking in credibility and legitimacy. In other studies, scale-related challenges of ignorance (e.g. national policies adversely constrain local policies), mismatch (e.g. differences in institutions governing resources and the biogeophysical scale of the resource), and plurality (e.g. failure to recognize differences in perceptions and values) have been identified as barriers to information exchange . The way discourses derived from ontologies are formulated into practical interventions is also particularly important for Indigenous peoples, whose knowledge, values and practices in natural resource management are currently undergoing significant change across the world.

The current dominance of market-based instruments (MBIs), such as payments for ecosystem services (PES) including REDD+, means that Indigenous peoples and other marginalized communities are at the forefront as 'implementers' of these environmental (and associated climate change mitigation) policies . In the context of PES and REDD+, this is not only through receiving funds to maintain sellable and tradable 'ecosystem services', but also to monitor, report and verify the status of the environment to fulfil state obligations to international funding organizations. Yet although MBIs espouse a range of creative and innovative solutions to environmental management, critically, their ideological foundations within a neoliberal agenda that promotes 'selling nature to save it', is in stark contradiction with Indigenous ontologies based on human–nonhuman–spiritual relationships . Recent studies indicate that in practice, MBIs not only escalate inequalities through privileging elites and intermediary organizations, but also fall short in 'permanence' as a result of lack of land tenure, corruption and the failure to defend community land rights from competing interests . In addition, financial incentives through MBIs have the potential to actually 'crowd out' pre-existing intrinsic environmental motivations and ethics, changing Indigenous value systems [16,17]. This all reveals an intention to assimilate Indigenous culture into the dominant and Western economic model, rather than supporting self-affirmation of Indigenous identity and autonomy .

Within this wider context of environmental governance and Indigenous practices, the use and management of fire continues to be a much debated and controversial topic. In the last decade, devastating wildfires have wreaked havoc on many tropical forest and savanna ecosystems, including those in the Amazon Basin. Carbon cycle studies of the Brazilian Amazon show that its current net carbon sink (net biome productivity, NBP) of +0.16 (ranging from 0.11 to +0.21) PgC year$^{"1}$, equivalent to 13.3% of global carbon emissions from land-use change for 2008, can be negated or reversed during drought years (NBP = "0.06 ("0.31 to +0.01) PgC year$^{"1}$), with forest fires likely to be the dominant flux (48.3% relative contribution) of carbon during extreme droughts . South America had the most important contribution to carbon emissions (37% between 1997 and 2009), particularly associated with post-clearing land use for pasture or soya bean plantations in the Brazilian Amazon, which represents a higher combustion completeness not compensated by regrowth on decadal time scales . Thus, climate change, deforestation and the expansion of agriculture are the major drivers for the increasing scale and frequency of wildfires in the region, and although national laws attempt to prevent and control the use of fire, the extensive burning of the landscape signifies a mismatch between fire policies and burning practices . With a growing recognition across the world that combatting landscape fires is not only ecologically, but also socially and economically unviable, drawing on traditional

Indigenous fire management could be a useful lens through which to find practical fire management solutions, but also lessons on how environmental governance could be structured and implemented more widely.

In this paper, we review existing data on Indigenous fire management, focusing on examples from tropical forest and savanna

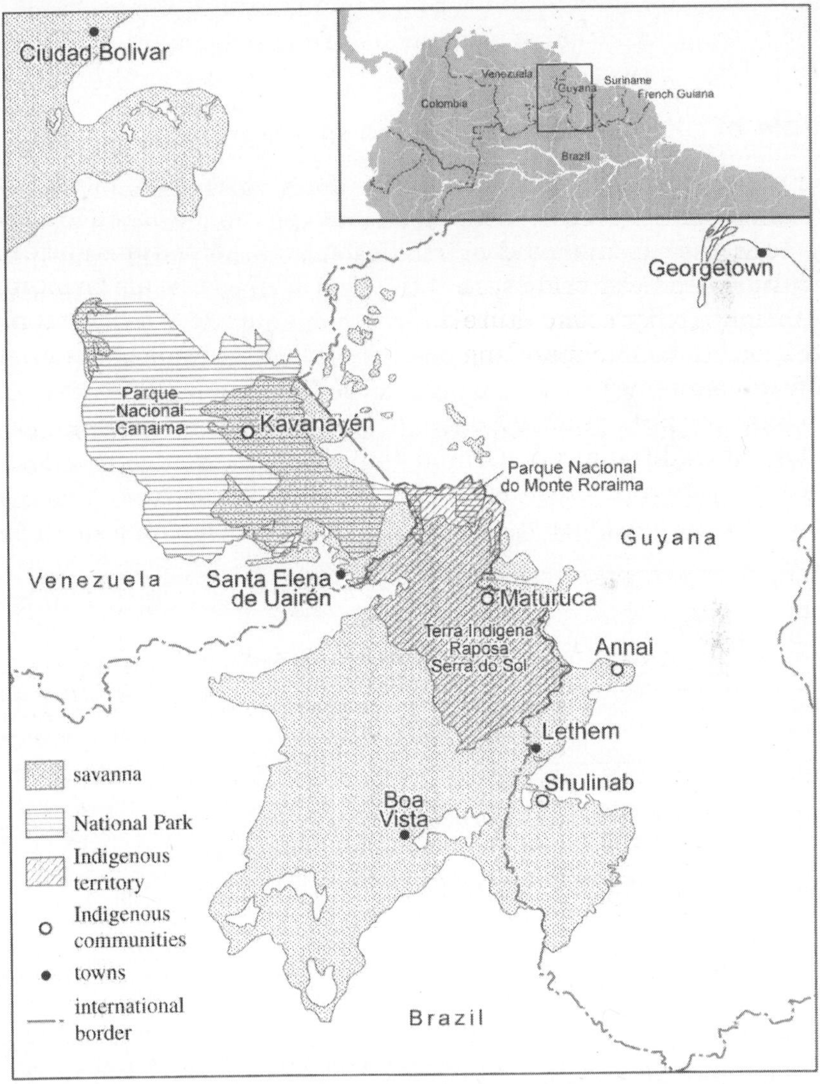

Figure 1. Map shows the geographical context for the Venezuelan, Brazilian and Guyanese Indigenous communities discussed in this article (kindly drawn by Jenny Kynaston).

ecosystems in Venezuela, Brazil and Guyana, countries in which we, as authors, have considerable fieldwork experience. We highlight the key attributes of fire management techniques stemming from Indigenous communities themselves, i.e. community owned solutions for fire management, and some of the challenges. We then go on to review institutional responses to fire management, and current approaches linked to climate change and MBIs. The aim of this paper is to reflect on the current situation of Indigenous fire management, and present possible ways forward.

2. State of play on Indigenous fire management

Fire is used by Indigenous peoples for a variety of purposes that have interconnected ecological, social and spiritual importance. These have been well documented by studies across the world, and include agricultural and pastoral use, hunting, gathering, fishing, stimulating vegetation growth and abundance, clearing vegetation, habitat protection, domestic use, and medicinal/healing and spiritual use. Huffman , assessing traditional fire knowledge documented in 35 studies, including accounts from 27 countries on six continents, proposes a typology of traditional fire use and associated knowledge based on the economic system of burning, or the agroecological type. These are: swidden—rotational farming through clearing and burning small forest

Table 1.Uses of fire by the Wapishana and Makushi people of the South Rupununi, Guyana (modified from Rodríguez *et al*.).

use	detail
domestic	cooking, heating and preserving food, warmth, light, cleaning around homes, burning rubbish, making and burning clay bricks
medicinal/healing and spiritual	preparing traditional medicines, healing (smoking out evil spirits), ceremonial practices, chasing away dangerous spirits or in some cases calling them (e.g. the rain spirits)
safety	cleaning paths, clearing around houses, chasing away dangerous animals (jaguars, snakes and mosquitoes)
animal husbandry	producing fresh green grass for grazing cattle, preventing cows from straying far away, finding lost animals such as pigs, getting rid of ticks, branding cows, herding cows
agriculture	opening new farms, fertilizing and cleaning (weeding) farms, chasing away ant pests
hunting and fishing	flushing out animals; as light
gathering natural resources	burning along swamps before cutting palm leaves to create space for drying leaves, smoking bees before collecting honey, stimulating certain trees to fruit
protection	preventing large fires entering forest-islands, farming areas, palm areas, homes, no-go zones; fighting large hazardous fires when approaching (fighting fire with fire); burning potentially dangerous overgrown swamps and savanna
communication	signals in hunting, grazing, emergencies
crafts	shaping crafts e.g. heating, bending and straightening arrow canes, bows, and fishing rods

patches; arborist—fire is used to maintain trees either in groves or individually, sometimes for stimulating fruit production and/or maintaining sacred sites; tame pasture—fires are used to maintain forage for domestic livestock in delineated pastures; and open native vegetation—fire is used in unconfined areas of expansive native vegetation for hunting, gathering, nomadic pastoralism, clearing travel routes, maintaining village sites, communication, etc.

Various studies show how fire use takes place at different times during the seasonal calendar in relation to particular livelihoods and resource management activities, and that this relates to fined-tuned Indigenous understandings of different environmental and climatic indicators. For example, many Indigenous groups throughout lowland South America traditionally have timed their agricultural cycles to the appearance of the Pleiades stars in the early evening and their movement across the sky`. The timings of fire are also aligned to phases of the moon. Importantly, the numerous uses of fire mean that burning is a relatively constant activity, particularly during the dry season, generally at low levels, thereby helping to prevent the build-up of flammable fuel and incidents of large-scale uncontrollable wildfires . Experimental studies of fire behaviour suggest that this patch mosaic burning not only reduces the occurrence of dangerous fires, but also increases spatial and temporal vegetation heterogeneity and biodiversity.

Indigenous fire management is effective in that it is an emergent property of a linked social–ecological system where Indigenous knowledge and culture, and associated livelihoods, are intimately interconnected with landscape management practices. For example, the Mebêngokrê (Kayapó) of the Capoto-Jarina in Mato Grosso, Brazil, use fire to hunt for land tortoises, which form part of an extended yearly traditional festival with implications for social processes, including courtship, community cohesion, youth initiation and intergenerational knowledge transfer (J. Mistry and A. Berardi 2014, personal observation). For the Pemón, the practice of Mayú—a system of mutual cooperation in the elaboration of large-scale tasks in traditional farming, e.g. cutting trees and burning the felled biomass—is not only essential for the survival of individuals, but also a social interaction facilitating the formation and establishment of social bonding and intergenerational knowledge transfer (B. Bilbao 2012, personal observation). Therefore, savanna and forest ecosystems are being protected within Indigenous lands not because they are being 'managed' in a direct and active way, but as the indirect outcome of a healthy social–ecological system, i.e. the outcome of practices that maintain social and ecological integrity, or what can be termed 'community owned solutions' [30, p. 10].

3. Challenges for Indigenous fire management

As stated above, Indigenous fire management, as with other Indigenous practices, is strongly tied to Indigenous knowledge and culture, as well as local governance structures and processes, which in turn are based on leadership and collective actions. For example, Mistry *et al.* explain how the social structures within the Krahô of Brazil determine the leaders or 'knowledge bearers' for fire, the Wakmejê, who decide on the different aspects of the fire burning regime during the dry season. Every morning at sunrise, the Wakmejê men meet at the centre of the village (the *ka*) to discuss the day's activities, and depending on the time of the season, the group decides on the course of action. Welch describes how, within the Xavante of Brazil, young mentors and elders help to encourage active learning, and entrust younger people to assume responsibility for their own acquisition and production of knowledge pertaining to the ecology of fire use, the burning calendar, and associated group hunting strategies and ceremonies.

However, the current status of traditional fire management within Indigenous communities can be associated with inter-related issues of a general loss of knowledge, a breakdown of social cohesion within communities, and conflicts (particularly ideological) between Indigenous and non-Indigenous stakeholders.

Among the Krahô, Brazil, for example, most burning occurred in groups and with the consensus of fire knowledge-bearing individuals. However, many younger Krahô men that had been influenced by outsiders, namely farmers of European descent, openly criticized burning during communal meetings, resulting in many early-season protective and resource-enhancing fire practices not being implemented . At the same time, because the fire practices of individuals were not under the scrutiny of the group, these continued to be used, mostly in the late dry season resulting in increasingly damaging fires that reinforced antagonisms against fire use. This was being exacerbated by the incidents of poachers coming into their land and using fire in the late dry season.

Evidence from the savannas of the South Rupununi, Guyana, suggests that landscapes may be subject to too many later dry-season fires and not enough savanna patch burning and forest-edge burning in the late wet season and early dry season as was traditionally done . The causes of this increase in dry-season fires were multiple and included changes in farming and hunting practices, an increase in the number of privately owned cattle, a lack of agreement amongst the different villages and with cattle ranchers over the use of fire for cattle grazing, and a general loss of fire knowledge by young people. For example, it was suggested that the increased burning by cowboys (*vaqueros*) in relation to pastoral and rodeo activities during the dry season was a result of young *vaqueros* no longer having the knowledge on how to use fire sustainably. At the same time, young Wapishana and Makushi and some community leaders were more critical about the use of fire as they had more regular contact with state natural resource management officials and environmental organizations that promoted anti-fire discourses . As with the Krahô, changing Indigenous values to focus on fire prevention and suppression could have the effect of making the problem worse.

In Roraima, Brazil, traditional practices in the use of fire for agriculture by Indigenous communities (Ingaricó, Makushi, Patamona, Taurepang, Wapishana and others) have been strongly criticised by governmental institutions, which have developed several initiatives to replace clearing and burning with the use of tractors under the slogan 'technology is white, not Indian' . Among Pemón communities of the Canaima National Park, Venezuela, many young people have been critical of traditional fire use, largely owing to a loss of traditional knowledge coupled with environmental education programmes focusing on fire control run by state resource managers from the National Parks Institute (INPARQUES) and the Caroni Electricity Company (formerly EDELCA, now CORPOELEC) . These intergenerational divides between young Pemón and elders have led to a decline in prescribed burning, which in turn has led to a build-up of flammable biomass and an increase in large-scale wildfires in the late dry season in some areas of the Gran Sabana [34,35].

According to Huffman , the status of traditional fire knowledge can be defined as: robust—fire systems that have persisted and continue to evolve over time, allowing for some changes in continuity but remaining essentially intact until today; declining—where traditional fire knowledge still exists within members of a given culture, but one in which demographic, economic, political, land use or other changes threaten its continued viability; rejuvenating—where active efforts are underway to both recover or to share traditional fire knowledge in landscapes in which traditional fire management was once the norm; or historical—where active fire management is no longer practiced and most traditional fire knowledge is largely historical, preserved in written, graphical or anecdotal accounts. Although we know that Indigenous groups from different localities have in-depth contextual knowledge on fire management, we also know that traditional fire knowledge is declining. Lehman , in the case of reviving burning practices of the Palawa of Tasmania, indicates that although there was enthusiasm for re-establishing early dry-season patch burning, these went out of control. Rejuvenating traditional fire practices in the short term is not easy or straightforward when settled and westernized Indigenous communities may have lost the depth of understanding developed over a thousand generations of living with the land.

There are concerns on how and to what extent existing knowledge is being or can be adapted to meet needs as local social–ecological systems change. For example, there are Indigenous observations of changing fire regimes in response to changing rainfall patterns—appearance of the Pleiades star has become or is becoming an unreliable indicator of the onset of the rainy season—with forests becoming drier and burning more easily and to a greater extent . In addition to larger forest and savanna fires during droughts, evidence from the South Rupununi, Guyana, indicates that changes in weather patterns could also be affecting the extent to which prescribed fire practices are carried out at the end of the rainy season in order to be prepared to manage fires in the dry season . Nevertheless, Bliege Bird *et al.* show that there are dynamic interactions between people and climate at the landscape level, with Aboriginal hunters buffering the impacts of climate variability with dramatically smaller but more numerous fires in dry cool conditions.

Another important but little discussed area is the impact of invasive species on fire regimes within Indigenous territories. Bardsley & Wiseman point to the growing evidence that the invasive plant species, buffel grass (*Cenchrus ciliaris*), has already reduced biodiversity in areas where it is well established, and is beginning to alter fire management practices in the Anangu territory in South Australia. In Brazil, we have seen the growing presence and spread of introduced pasture and invasive species such as *Brachiaria decumbens* within Indigenous territories, and subsequent higher-intensity fire events leading to forest degradation (J. Mistry and A. Berardi 2015, personal observations). Climate change will only exacerbate these fire and invasive species interactions by, for example, increasing dry biomass loads of these fast-growing, non-native species, especially during La Niña years of high rainfall, especially in areas where seed of pasture grass is already abundant .

4. Institutional approaches to fire management

Until recently, the paradigm of 'zero fire' was the norm in many areas of the world, including Venezuela, Brazil and Guyana. All fires were seen as a threat to biodiversity conservation and natural resource management, and the institutional response was to suppress the intentional lighting of fires; and when these did occur, to 'fight' or 'combat' fire through policies supporting fire suppression/protection and firefighting. Colonial explorers, missionaries and naturalists were the first to advocate a negative narrative of traditional fire use as degrading and harmful to the environment, exacerbated in recent years by scientists, the media, politicians and tourists . In the Gran Sabana, Canaima National Park, Venezuela, for example, the derogatory phrase '*Pemones los quemÌones*' (crudely translated as 'Pemón the pyromaniacs') has generated considerable conflict over fire management between state resource management actors and Indigenous peoples, reflecting wider Indigenous struggles over territorial land claims and self-determination . Similarly, the conservation discourse in Brazil which has historically believed that all anthropogenic burning, including Indigenous, is destructive, has also been widely adopted by powerful interest groups such as the '*ruralistas*' (bloc of large pro-agrobusiness landowners) as part of the political narrative contesting Indigenous rights to land . This is in the context of international climate change mitigation strategies such as REDD+ in which fire's association with deforestation, whether small-scale through Indigenous shifting cultivation, or large-scale through logging and agricultural expansion, is under scrutiny for compromising carbon permanence and undermining the potential of sustainable forest management .

(a) Venezuelan policy environment

Venezuela has one of the most progressive Indigenous rights regimes in South America. It is the only state in South America that officially recognizes, within its constitution, Indigenous peoples' rights to maintain their own production practices, and protects collective intellectual property of knowledge, technologies and innovations . Specific legislation focusing on Indigenous communities, such as the *Ley Orgánica de Pueblos y Comunidades Indígenas* , has also strengthened Indigenous rights to genetic resources and ancestral knowledge. Despite the progressive constitution and regulations, and the relative protection of Indigenous rights in national parks, the use of fire is still heavily restricted and combated (Art. 65, *Ley Penal del Ambiente*) . National parks have priority measures of protection and the National Institute of Parks (INPARQUES) is the highest administrative authority in charge of their management and conservation. INPARQUES, created in 1978, is an autonomous body ascribed to the Ministry of Power of the People for Eco-socialism and Water (MINEA, formerly Ministry of Power of the People for the Environment). Thus, the environmental authorities are in charge of the prevention and control of fires.

In 1978, a Programme for Prevention and Protection Against Forests was created to impart knowledge, experience and skills for training forest rangers, but it was only in 2001 during the Presidency of Hugo Chavez, that the Law of Fire-Fighters and Civil Emergency Management (*Ley de Cuerpos de Bomberos y Bomberas y Administración de Emergencias de Carácter Civil*) , was enacted, promoting, some years later, the graduation of the first professional forest firefighters in the country. This law also gave rise to the Unified National Command against Forest Fires, to coordinate the different entities—MINEA, INPARQUES, the Bolivarian National Guard and Civil Protection, regional and local bodies—during the dry season to prevent and mitigate fires in the country.

One of the firefighting programmes in Venezuela with the longest trajectory, and better endowed in terms of equipment and infrastructure (helicopters, aeroplanes, landing strips, guardhouses, etc.), is the Control of Wildfire Programme (PCIV), implemented by the Initial Attack Brigade Carlos Todd of CORPOELEC (the regional hydroelectricity company). The PCIV was created in 1981 to prevent, detect and fight wildfires in order to protect the headwaters of the Caroní River which supplies the Guri Reservoir and the *Central Hidroeléctrica Simón Bolívar*, where 70% of the country's hydroelectric power is generated. 21 000 km^2 of this highly protected area is located in the Gran Sabana and includes the eastern sector of Canaima National Park . Although members of the PCIV brigade are Indigenous Pemón, Indigenous knowledge has not been used in firefighting practices, and fire exclusion has been the official fire policy adopted in the Park. In spite of carrying out expensive and enormous fire suppression efforts, on average, only 13% of total fires are combated owing to the high number of fires over a large area . Moreover, the Pemón use fire in their daily activities, constituting an essential part of their cultural identity and contributing to their livelihood needs . Thus, the conservation policies undertaken in the past few decades in the Canaima National Park (reflecting the national context) have not only been based on the exclusion of fire, but have also largely ignored the perceptions, expectations and knowledge of its inhabitants, which has led to serious conflict between the Pemón people and government agencies.

(b) Brazilian policy environment

In Brazil, the Forest Code (first issued in 1934 and most recently revised in 2012) was the first piece of legislation to make burning without specified precautions illegal (Art. 22) . In 1989, decree No. 97,635 regulated article 27 of the Forest Code, and created the National System of Prevention and Control of Forest Fires (PREVFOGO). The aim of PREVFOGO was to support the monitoring, prevention and combat of forest fires, and to develop and disseminate technical training and education in fire control. However, although fire was mentioned in various environmental governance laws, it did not appear in specific regulations until 1998, following the large forest fires in Roraima induced by the 1997–1998 El Niño southern oscillation. The national and international outcry from these fires prompted the federal government to create decree No. 2661 redefining the role of PREVFOGO and, for the first time, establishing detailed procedures on the use of prescribed fires. In addition, a series of new fire policies and management initiatives were introduced, including the Programme for Prevention and Control of Fires in the Brazilian Amazon Forest (PROARCO) with the main goal of controlling and preventing large-scale wildfires in the Brazilian Amazon.

Since then, fire management has been mostly subsumed within the climate change agenda, through the 2009 National Climate Change Policy and its programmes, which include the Programme for Combating Deforestation in the Amazon (PPCDAm), the Programme for Combating Deforestation and Forest Fires in the Cerrado (PPCerrado), state-level plans, and the Low-Carbon Agriculture Plan (ABC) . For example, under PPCerrado, the Projeto Cerrado/Jalapão aims to improve the prevention and control of irregular burning and forest fires in the Jalapão region, thus contributing to the maintenance of the Cerrado as a globally relevant carbon sink . Activities include training, awareness raising and environmental education, implementation of demonstrations of alternatives to the use of fire, and the strengthening of state committees for preventing and fighting fires. In addition, the project has been testing an integrated fire management (*manejo integrado do fogo*, MIF) approach based on prescribed early dry-season burning to create patchy landscapes and reduce the probability of large late–dry season fires. MIF was applied in 2014 within three conservation areas—the Jalapão State Park (Tocantins), the Ecological Station of Serra Geral do Tocantins

(Tocantins) and the Chapada das Mesas National Park (Maranhão). The main objective was to inform controlled burning activities by observations from satellite sensors indicating the degree of biomass desiccation across the landscape, accompanied by simultaneous biomass and fire measurements in the field. The approach, in this case, was to identify, in real-time, emerging small patches of drying vegetation appropriate for burning, which would not spread easily into the surrounding more humid vegetation. Repeated use of this technique throughout the dry season would result in a mosaic of small burn scars, which would result in a reduction of all combustible biomass, while preventing larger and destructive wildfires from occurring. In 2015, the MIF was applied in the Xerente Indigenous territory (Tocantins) following a study of traditional fire knowledge with elders of the community to develop a fire calendar and prescribed-burn planning using remotely sensed imagery combined with geoprocessing tools such as spectral angle mapping.

The results of MIF are still forthcoming, but they do show us a change in approach within Brazilian institutions. Not only is there a move away from categorizing all fire as 'bad'; there is also a recognition that Indigenous fire knowledge is a valid form of knowledge that could inform policy-making. Nevertheless, efforts to actively involve Indigenous people in fire management have to date mostly been in the form of fire brigades. There are currently 34 Indigenous brigades employed during the dry-season period to help combat fires.

(c) Guyanese policy environment

In Guyana, fire has been inextricably linked with forest management and timber extractive activities. As a driver of deforestation, fire has been implicated in the recent REDD+ activities through which Guyana aims to implement its low-carbon development strategy; a national plan to reorient Guyana's economy on to a low-carbon path . The latest Readiness Preparation Proposal for REDD+ [56, p. 62] outlines national-level activities to be conducted to achieve readiness for the implementation of a forest carbon financing mechanism, and includes the formation of a national forest fire management strategy. Rodríguez *et al.* report on the launch by the Guyana Forestry Commission of a pilot dry-season fire-monitoring programme in various communities of the South Rupununi in November 2010, and a recent aerial fire assessment was undertaken in the same locations by the Ministry of Natural Resources and the Environment . However, there has been little evidence that any fire management policy has been developed to date, and there seems to be no recognition within policy-making of traditional fire use in savannas by Indigenous groups.

5. Emerging market-based instruments integrating Indigenous fire management

The case of Australia is heralded as a progressive example in which Indigenous fire knowledge is being incorporated into market-based instruments for fire management that includes goals for carbon cycling and reducing GHG emissions. Savanna fire management in northern Australia, where early dry-season prescribed burns are used to prevent late dry-season wildfires, is an approved offset methodology under Australia's Carbon Farming Initiative (CFI) . Australian Carbon Credit Units generated through the CFI can contribute to meeting Australia's commitments under international agreements to reduce GHG emissions.

For example, the Western Arnhem Land Fire Abatement (WALFA) and the Central Arnhem Land Fire Abatement (CALFA) projects are PES schemes that aim to reduce GHG emissions in the atmosphere while creating employment opportunities for Indigenous people in remote regions. WALFA is mainly funded by a subsidiary of the multinational ConocoPhillips, which aims to partially offset its emissions of GHGs from a liquefied natural gas plant in Darwin (100 000 tonnes of CO_2 equivalent annually) against AU$1 million a year for 17 years. This funding supports several Indigenous ranger groups located in western Arnhem Land, which work in collaboration with traditional owners of the region as well as fire ecologists. The multinational is thus seeking to reduce its carbon footprint and abate its GHG emissions not by limiting its own production and release of such gases, but by funding fire management programmes that are led by and benefit Indigenous people. The CALFA project is organized along the same lines as WALFA (though initially supported through public subsidies with private funding expected over the longer term), and involves several Indigenous ranger groups working in central and southeastern Arnhem Land.

The perceived success of the Australian experience of savanna fire management, particularly the WALFA case study and model (reducing GHG emissions by 30%) , has led to Australia funding an initiative through its aid budget to promote the international applicability of the fire management methodology and experience.

However, Petty *et al.* show how these new emissions-reducing programmes run the risk of further marginalizing Indigenous people. Inherent in the nature of institutionalized management programmes is to replace the complexity and contingency of Indigenous fire management with standardized goals, while treating Indigenous people as workers executing plans developed by others rather than as genuine partners. They suggest that the funding model for the WALFA project creates an incentive to focus on following a complex emissions-accounting methodology that '...ties the practice of burning very tightly to the accounting of burning. This forces a close registering and recording of burning, whose metrics are then tied to an external scheme, and represents a dramatic shift in the character of Aboriginal fire management, which is rooted in place based knowledge, dynamic decision making, and attention to unique seasonal changes in vegetation' [62, p. 157]. The approach also creates a subgroup of Indigenous rangers, who are tasked to carry out systematic fire management in the early dry season at the behest of the institutions they represent. However, they are criticized by the community members for not having in-depth knowledge and applying too much fire carelessly, while at the same time they are excluded from using the technocratic fire management resources for lack of training and skills.

Indeed, in many countries, there is a reliance on technological applications as a solution to fire management, with a clear (and dominant) discourse aligned with geospatial technology, where voice and power over decision-making is in the hands of those with the technology (e.g. scientists, governments) rather than those without (e.g. local farmers, Indigenous peoples) . In a review of fire studies in tropical humid forest areas, Carmenta *et al.* found that remote sensing techniques for detecting fire were favoured by resource/park managers and policy-makers because of their replicability and representation of a seemingly objective reality. Indigenous communities are excluded from this technocratic approach in that the overly simplistic remotely sensed 'reality' does not correspond to the multidimensional (spiritual, social, ecological) experiences as perceived by Indigenous people. Most importantly, the institutionalization of Indigenous fire management, and its scientific and technocratic discourse strongly privileges one particular aspect of Indigenous fire management: early dry-season burning to protect against late dry-season burning . This fails to recognize that Indigenous fire management is characterized by regular and sometimes opportunistic burning throughout the dry season linked to various social, ecological and spiritual purposes, which can buffer the impacts of climate variability and produce habitat mosaics that support landscape biodiversity.

6. A 'case' for supporting Indigenous fire practices within government fire management policy

Our review of the current literature on Indigenous fire management highlights the following:

- Fire is an integral component of savanna and forest landscapes worldwide.—
- Some— Indigenous people have detailed knowledge on fire management that could help prevent large-scale and destructive wildfires and associated land-use change.
- Although there are— examples of 'robust' Indigenous fire management, it is 'declining' in many areas leading to the increase of inappropriate and damaging fire occurrence. This is a result of the interplay between loss of Indigenous knowledge, a breakdown of social relationships and cohesion, and conflicts (particularly of worldviews) between Indigenous and non-Indigenous stakeholders.
- Incentive-— and market-based approaches run the risk of focusing on the management of isolated issues, e.g. carbon storage, without adequate understanding of inter-relationships and interdependencies and thereby simplifying the complexities of Indigenous fire management. Associated perverse financial incentives could also 'crowd out' community cohesion and intrinsic values that are essential for effective Indigenous fire management. Requirements (from donors, states) for monitoring and reporting could remove control of fire management from Indigenous communities.
- Current fire policies, and— associated institutional structures, strongly focus on suppression and firefighting, with Indigenous participation mostly in the form of institutionalized and 'professionalized' rangers/brigades. This runs the risk of marginalizing the wider community, disrupting traditional modes of knowledge transfer and therefore loss of Indigenous knowledge, and conflicts between different sectors of the community.
- Attempts— to institutionalize Indigenous fire management have focused on the use of early dry-season fires at the expense of the complex and sometimes continuous burning throughout the dry season and in the wet season.

We can see from this that Indigenous fire management is being incorporated into policies through already established and clearly defined government schemes; *disincentivizing*, command-and-control methods of firefighting through the creation of Indigenous fire brigades, and *incentivising* approaches focused on prescribed early dry-season burning. Our contention is that although firefighting and early prescribed burning are necessary as part of an overall fire management strategy, there needs to be *enabling* policies that focus on legitimizing and strengthening Indigenous fire management as a community owned solution. Critically, as community owned fire management is intricately linked with Indigenous survival strategies, so too must firefighting and prescribed burning be grounded in local social–ecological systems. We believe it is necessary to define long-term actions to support the integrated functioning and survival of Indigenous communities as a whole, rather than focusing on isolated issues (e.g. carbon retention) or benefits for some individuals (e.g. hiring Indigenous firefighters).

The 'State of the World's Indigenous Peoples' report provides insights into the differences in perspectives between worldviews. John Bamba, an Indigenous Dayak from Kalimantan, Indonesia, summarizes the underlying principles for living a good life based on the Dayak's traditional cultural values as sustainability, collectivity, naturality, spirituality, process orientation, domesticity and locality. These are contrasted with prevailing modern values—productivity, individualism, technology, rationality, efficiency, commercialism and globalization—that have become predominant principles in present-day social and economic development, and can undermine a balanced human–nature relationship. What we want to do is not promote one over the other, but encourage decision-makers to engage with, and appreciate, Indigenous perspectives and worldviews on fire management. Community owned solutions acknowledges collectivity, spirituality, process orientation and locality, whereas many expert-led fire management interventions often result in promoting individualism, ethnocentrism, rationality, efficiency, commercialism and globalization. The question we raise is this: can the 'community owned solutions' approach be the mechanism through which Indigenous perspectives can be represented within fire management?

Our aim now, focusing on Venezuela, Brazil and Guyana, is to bring together all parties involved in the fire decision-making process—Indigenous people, policy-makers, scientists, park managers and private landowners—to share perspectives, and respect and support Indigenous survival strategies within which fire management is embedded, while at the same time contemplating the needs and constraints existing in the system to be managed, and developing action plans to make change. This process to integrate Indigenous fire practices within government fire management policy focuses on 'enabling' policies centred on legitimizing and strengthening Indigenous fire management as a community owned solution in order to promote the empowerment of Indigenous communities and their active participation in decision-making. Actions have to be aimed at encouraging Indigenous communities more autonomy with respect to implementing policies, including the leadership and funding of fire management programmes.

Indigenous fire management

Fire is perceived through the lens of culture, and is rooted in mythology and religion. In Ancient Greek mythology, the Titan Prometheus took fire from the Gods, and Indigenous cultures from the Pacific Northwest tell of fire being stolen by coyote, beaver, and dog, and then gifted to humans (Judson 1912). Western religions cast fire as being pagan, destructive, and even apocalyptic, shaping modern Western perspectives on fire, and fire management (Pyne 2016). Despite the different narratives, fire is central to human development, and has transformed our societies and landscapes.

Paleoecological and oral evidence shows Indigenous societies have used fire to manage landscapes regularly over millennia, and many groups still actively do this today. Many fire-dependent ecosystems are reliant on anthropogenic fire, and fire is used to create food security, termed "fire-stick farming" in Australia (Jones 2012). Yet, a paradox exists, where the global land area burned has declined over the last two decades because of factors such as agricultural expansion and fire suppression (Andela et al. 2017), while the incidence of destructive fires is increasing because of factors like climate change and mismanagement. The world is not seeing more fire, but is experiencing the wrong type of fire, in the wrong place, at the wrong time (Pyne 2016). Controlled burning to mitigate uncontrolled fires is critical in fire-dependent ecosystems. However, significant legal, political, and attitudinal barriers constrain this practice. Indigenous fire management (IFM) is increasingly touted as a way to bring fire back to landscapes (Russell-Smith et al. 2013, Mistry et al. 2016, Lake and Christianson 2019, Moura et al. 2019), but what is IFM?

Indigenous peoples across the world manage their lands with fire, guided by their knowledge, practices, lore, and customs. IFM is the proactive use of fire to achieve multiple and complex landscape-level objectives, such as cleaning the landscape, mitigating destructive wildfires, and reducing greenhouse gas emissions; ceremony; promoting biodiversity and food security; and generating sustainable livelihoods. We examine scholarly literature on IFM and its accompanying ontological (the nature of reality), epistemological (understanding reality), and methodological (ways for gaining knowledge about reality) complexities. Using a conceptual framework analysis approach, a grounded theory methodology, we generate a conceptual model of IFM from scholarly literature, which illustrates its main concepts and their interrelationships as expressed in literature. We acknowledge that much of the scholarly literature is from a Western perspective, which is a significant limitation to this literature review. This paper offers a call to action to include Indigenous knowledge and voices on this important topic. The conceptual model developed from the literature can guide research and practice on this multidisciplinary phenomenon.

CONTEXT

Fire suppression, reflected in the figure of Smokey the Bear in North America, has dominated landscapes across the globe during the 20th century . Fire is to be avoided and if unavoidable, extinguished to protect life and property (Pyne 1982). In a warmer world, uncontrolled wildfire remains a constant danger in fire prone areas, or "pyroscapes," with catastrophic megafires threatening ecosystems, infrastructure, and human populations. Indigenous peoples of pyroscapes often possess knowledge of fire and its use on these landscapes (Huffman 2013). Indigenous burning practices have notably shaped these pyroscapes over time (Bowman et al. 2011, Steen-Adams et al. 2019).

Historically, as European colonizers expanded across landscapes, IFM regimes were drastically altered (or quashed), which had consequences for the relationship between humans and fire, as well as negative impacts on landscape health. IFM has declined since colonization, but it has continued in many areas and is being reactivated despite political barriers and conflicting worldviews with non-Indigenous stakeholders (Mistry et al. 2016). Landscapes have since been altered in fundamental ways by deforestation, agricultural development, grazing, invasive species, forest encroachment, urbanization, and climate change. In many cases, Indigenous fire knowledge and practice needs to be adapted to current and future conditions.

Recent record-breaking fires around the world have led to calls for decentralized and proactive fire management beyond fire suppression.

A report by Abbott and Chapman (2018) in British Columbia, Canada called for the prioritization of proactive fire prevention, prescribed burning, and better fire management coordination with First Nations (Indigenous peoples). In the United States, a Forests and Rangelands (2014) government report recommended wildfire mitigation through risk reduction strategies and landscape restoration. An Australian CSIRO (2009) report proposed an adaptive fire management approach in order to accommodate climate change and biodiversity. There were also public calls for IFM in Australia after the 2019 fire season.

The United Nations Sendai Framework for Disaster Risk Reduction 2015-2030, outlined that "Indigenous peoples, through their experience and traditional knowledge, provide an important contribution to the development and implementation of plans and mechanisms" for disaster risk reduction (United Nations Office for Disaster Risk Reduction 2015:23). The academy has also advocated for more Indigenous involvement in fire management. Indigenous fire knowledge and practice can generate multiple social, economic, cultural, and environmental benefits (Northern Australia Environmental Resources Hub 2016) and IFM programs have been implemented in Australia, Canada, the United States, Brazil, and recently in Botswana. These programs involve collaborations between Indigenous and Western knowledge systems, and are embedded in, and shaped by the institutional context. Examining how IFM plays out within these collaborative processes and the institutional context is critical to understanding the development of IFM, and how it is applied as an alternative to fire suppression in specific areas.

Institutional and policy context

IFM was prohibited (and often displaced) over large areas, particularly where Indigenous peoples were removed from their lands and resources, and forbidden from practicing their culture and language (Nikolakis 2019, Nikolakis and Nelson 2019, Nikolakis and Hotte 2020, Nikolakis et al. 2020). The emergence of IFM as a "legitimate" alternative to fire suppression involves a reactivation of Indigenous knowledge and enhanced participation in land governance.

In countries like Canada, Australia, and the United States, subnational governments (states or provinces) typically have jurisdiction over fire management. On Indigenous lands, the federal or relevant Indigenous government is typically responsible for fire management. There can be a clash of values and goals for fire management, with subnational governments adopting a fire suppression approach, driven by Western knowledge and practices. Whereas Indigenous worldviews, knowledge, and practices call for a more place-based and holistic approach using fire as a tool. In practice, it is common that fire management regimes neither match the ecological fire regime, nor do they draw from all available knowledge sources; this leads to negative ecological and social outcomes.

Indigenous fire management goals are diverse, ranging from hazard reduction (Lewis et al. 2018), to maintaining biodiversity and cultural values, like protecting ceremonial areas, to more localized cultural burning practices throughout the year and enhancing livelihoods. For subnational governments, the goals are calibrated to land tenure, such as industrial forestry, rangeland, tourism, and conservation areas. In conservation areas, the fire-related goals are to maintain specific ecological values and to reduce wildfire risk, often achieved through prescribed burns. Efforts to integrate these different values and goals through comanagement approaches are well documented, and require parties to respect value differences, and to create rules about how these forms of knowledge and practice are integrated.

Barriers to implementing IFM include public perceptions of safety, air quality concerns, and other political complications (Marsden-Smedley and Kirkpatrick 2000). There are actor networks and advocacy coalitions advocating for IFM in policy processes, as a legitimate alternative to fire suppression (Diver 2016). Understanding the key concepts of IFM, and the interactions between these, is crucial to constructing a theory of IFM within the policy process.

CONCEPTUAL BACKGROUND

Literature

Early IFM literature took an etic approach to understanding precontact Indigenous fire regimes (Pyne 1982, Boyd 1999), studying communities from the outside. Recent literature increasingly takes an emic approach, as scholars build relationships with Indigenous communities, or communities produce their own literature, with the aim of exploring how Indigenous knowledge and practices interact with (or outside of) contemporary fire management. This often occurs in protected areas or on Indigenous lands. Most studies on contemporary IFM are at a local scale and measure cause and effect, governance, and the outcomes of IFM programs (particularly in Australia). Literature reviews have taken a social science lens (Christianson 2015) and a natural science lens focuses on prescribed burning. Comprehensive studies have summarized the historical impact of humans on global fire regimes (Bowman et al. 2011), as well as the commonalities of Indigenous fire knowledge worldwide (Huffman 2013). However, there is a lack of global interdisciplinary synthesis of IFM.

Conceptual drivers

Ontology, epistemology, and methodology are crucial to understanding IFM. Ontology refers to the nature of reality, which is socially constructed through cultural symbols, ideas, belief systems, and narratives (Potter 1996). Reality can be reframed through shock events, or through actors reorienting how reality is perceived through learning. Epistemology refers to understanding "reality," determining what constitutes knowledge, distinguishing "truth" from simple opinion, and choosing what is worth knowing. Societies with diverse cultures, politics, and histories will have different ways of evaluating knowledge and determining truth. Methodology is concerned with the methods, procedures, tools, and techniques for gaining knowledge. Methodology offers a justification and rationale for why certain procedures and tools are applied to investigate a given problem (Singh 2015). A methodology determines the approach for inquiry, and is guided by a researcher's ontological and epistemological position as they attempt to understand reality.

There are different knowledge systems for understanding the reality of fire and its management. Indigenous and Western science knowledge systems have differing (and sometimes common) epistemologies for how knowledge is uncovered or understood. There has been increasing focus on integrating Western and Indigenous knowledge to better understand and manage ecosystems and natural resources (Bohensky and Maru 2011). Western scientific paradigms create knowledge through critical reasoning, replicable methodologies, peer review, and written documentation, with ecological science increasingly focused on holistic systems thinking, resilience theory, and social-ecological systems. Indigenous knowledge systems and science are often more experiential, localized, holistic, and are typically passed down orally through generations. Both seek to understand phenomena by observing cause and effect, but the epistemological differences present challenges for collaborative approaches. Traditional knowledge holders may be hesitant to collaborate with the etic approach of Western science, uneven power dynamics, and skepticism about Indigenous knowledge. IFM in practice has mitigated these concerns by drawing from both Indigenous and Western scientific paradigms, and prioritizing Indigenous community participation in the application of knowledge. This, in turn, allows methodologies to be created for finding solutions to fire management problems at broader scales.

There are various risks, perceived and real, for IFM as a methodology. Christensen (2003) and Stankey et al. (2003) document that mainstream fire management is generally risk averse. Decision makers avoid short-term risk in lieu of potentially long-term ecosystem benefits (Ryan et al. 2013). Fire management actors use mental shortcuts that introduce systemic bias and create a mismatch between objectives and practices (Maguire and Albright 2005). Often the precautionary principle is used, whereby the absence of fire is framed as a "safe alternative" to "risky" prescribed burning practices. Status quo bias is also prominent, where land managers feel less responsibility for the negative effects of "doing nothing," i.e., wildfire caused by climate change, than for the negative effects of deliberate interventions, i.e., damage from an escaped prescribed burn. Regret theory also plays a role, whereby land managers minimize potential losses rather than maximize the net value of a decision (Maguire and Albright 2005). These mental shortcuts inhibit innovative fire management policies and practices.

IFM as a solution

Locating IFM within broader environmental decision making brings insight to context. Figure 1 illustrates that environmental problems, their nature, and their recognition, are rooted in decision makers' ontologies. Once a problem is acknowledged, a risk-assessment is made and potential interventions are evaluated, drawing on existing knowledge and technical tools. The portfolio of interventions for fire management are evaluated, be it fire suppression (status quo), prescribed burning, mechanical thinning, or a mixture of these. Where an intervention has previously been effective, it becomes the default option. Where the intervention has been ineffective, or the problem is unknown, then a space is created for interventions outside fire suppression (like IFM). If the solution is effective, we hypothesize this feedback will reshape existing ontologies to legitimate the innovation.

METHODS

To better understand emergent and multidisciplinary phenomena,

Jabareen (2009) developed a "conceptual framework analysis" (CFA) approach, a grounded theory technique that builds a conceptual framework of "interlinked concepts that together provide a comprehensive understanding of a phenomenon" (Jabareen 2009:51). The goal of the analysis presented in this paper is to utilize the CFA approach to organize key concepts of IFM in ways that demonstrate interrelationships, as well as their ontological, epistemological, and methodological dimensions. The intention is to produce an illustrative framework of the multidisciplinary phenomenon in question, that contributes to its overall understanding. We employ Jabareen's eight-phase qualitative analysis to build a conceptual framework, which together compose a process of theorization. The theories in interdisciplinary literature form the data analyzed to build the conceptual framework, emphasizing a continuous interplay between data collection and analysis.

Conceptual framework analysis

The first phase involved a literature review of IFM, drawing on scholarly literature from ecology, fire science, community development, economics, and policy. This review identified 349 papers on IFM, and 72 were selected for deeper review, based on the following criteria: they were published after 2000 (to reflect contemporary rights and knowledge), and published in peer-reviewed journals. The second phase involved extensive reading and categorization of data, where literature was organized by discipline and relative importance within each discipline (by citations).

The third phase involved the identification of concepts within literature. The main findings and recommendations from each study were collated and grouped into overall themes, and from these groupings we documented common concepts. The fourth phase involved deconstructing each concept by identifying its main attributes and characteristics, and then organizing these concepts according to their ontological, epistemological, or methodological role. In the fifth phase, concepts were integrated, where they were reconsidered, combined, or dispensed with. The concepts were then synthesized into a theoretical framework (here presented as a contextual conceptual model) during the sixth phase.

The seventh phase involved validating the conceptual model to ensure it represents a reasonable theory for IFM from literature. In the eighth and final phase, the conceptual model was shared with academics and practitioners, and their feedback was incorporated where appropriate.

Fifteen academics and fire practitioners were interviewed, including four academics from the U.S. (one Indigenous academic), four from Australia (one Indigenous practitioner), five from Canada (one Indigenous academic and practitioner), and two academics with work focused on Brazil and southern Africa.

RESULTS

A conceptual analysis identified five key concepts in the 72 articles. (1) Distinct ontologies: the ontology of fire is socially constructed and perceived differently by Western and Indigenous worldviews; (2) Governance: IFM often involves collaboration between Indigenous and Western knowledge systems, and occurs within a contested institutional context. (3) Revitalizing Indigenous knowledge: Indigenous knowledge and practices are rooted in an experiential epistemology where the landscape drives decision making. (4) Cobenefits: IFM can generate a range of ecological, social, economic, and cultural benefits. (5) Desired state: IFM can achieve a desired ecosystem state of resilience.

Concept #1: distinct ontologies: fire ontologies are distinct in Western and Indigenous worldviews

Relationally distinct fire ontologies exist: fire is either "good" or "bad," a "tool" or "hazard," "destructive" or "generative," and these views interact within existing institutional frameworks to shape the dominant epistemologies and methodologies for fire management.

Western fire ontologies have evolved over time. Heraclitus, of ancient Greece, believed all things were manifestations of fire (Wheelwright 1974), and later Aristotle classified fire as one of four elemental substances (Gomes 2018). In the 17th century, phlogiston theory reasoned all combustible materials contained the element "phlogiston," released into the air when something burned. This theory was gradually superseded by our current scientific knowledge that fire is the combustion of fuels through the process of an exothermic reduction-oxidation reaction. As Western fire knowledge has evolved, so too has people's relationship with fire. Fire was a central aspect of domestic and daily routine, but following the Enlightenment and advances in technology, fire became less integrated into people's lives (Pyne 2016). Modernity has seen fire replaced by electricity for some, and the condemnation of fire by agronomists and foresters has reduced its use in land management and affirmed perceptions of fire as a destructive force to be suppressed.

In contrast, for many Indigenous peoples, the use of fire is based on a reciprocal relationship between the individual and the universe. Many Indigenous worldviews believe the Earth is a generative, living being, and that humans and all living beings are interconnected (Kidwell et al. 2002). As oral knowledge is important to Indigenous societies (Berkes 1999), fire knowledge, of when and how to use fire on the landscape, is often found in people's stories. These stories set out various "laws of the land" (Eriksen and Hankins 2014), for the responsibility of fire as a management tool and often incorporate the intrinsic value of nature, the potential for the landscape to be alive or sentient, and the spiritual dimensions of the land. Fire is viewed by many Indigenous societies as a source of renewal, and burning cleans the landscape.

A clash of ontologies plays out in fire management in important ways. Colonizers brought European forest and fire management to areas they colonized and settled, despite many landscapes being fire-prone and fire-adapted (Pyne 2016). Colonization and fire suppression had drastic impacts on Indigenous people's ability to exercise their responsibilities to the land. It also changed these landscapes in important ways, as forests encroached on grasslands and fuel loads built up, enhancing the risk of devastating wildfire. IFM can not only mitigate risks, but can also contribute to the decolonization of fire management.

A significant ontological barrier to IFM is the idea of "traditional," where Western thinking frames traditional practices as those predating contact, and suggesting these are static (Pierotti 2018). Government agencies can simply reject IFM as an artefact with no place in modern landscapes (Mathews 2005). However, Indigenous cultures, worldviews, and knowledge systems are dynamic and adaptive, and can be applied to diverse settings and contexts. These distinct fire ontologies explain the divide in epistemologies and methodologies for fire management in colonized landscapes; the questions remain whether IFM can and should drive a convergence in ontologies.

Concept #2: governance: IFM is a collaboration influenced by political and power dynamics

IFM takes place in a complex governance environment. IFM programs are often collaborative and draw from multiple bodies of knowledge and normative views of what should (and should not) be done to the landscape. Learnings across Indigenous and Western knowledge systems can foster new fire knowledge and goals, one prominent example being a community savanna-burning methodology in northern Australia. However, in practice, the broader institutional framework may entrench uneven power dynamics that influence knowledge production (Ballet et al. 2007, Hunt 2014). Mainstream actors and agencies may remain ignorant to the potential of IFM (Mathews 2005) and Indigenous peoples may determine it unacceptable to integrate their fire knowledge with, or having this validated by, Western science. As Miller et al. (2010:2298) argued, "Indigenous knowledge is inextricably linked to the rights of indigenous people to make choices, to preserve and recreate knowledge through continuing engagement and practice on the land."

Tenure systems, land management bureaucracy, and vested economic interests impede Indigenous people's access to environmental governance and climate adaptation strategies. On top of this, strict antifire laws often prohibit Indigenous communities from burning, which puts IFM at risk. International commitments made by states can either impede or complement IFM, further contributing to complexity in the governance environment. For example, commitments to reduce greenhouse gas emissions in the Kyoto Protocol enabled savanna burning as an accountable activity, allowing IFM to be integrated into carbon markets in Australia (see Russell-Smith et al. 2013), while also drawing criticism for how such market integration can institutionalize Indigenous knowledge. Comanagement involving Indigenous peoples and the state is one way to bridge epistemologies and methodologies for land governance, particularly where this respects the authority and sovereignty of the groups involved. The reorientation of authority through comanagement, and more broadly self-governance, may facilitate IFM at a larger scale.

A lack of funding and institutional support also impede IFM. IFM is embedded in a complex governance environment, at various scales. As Mistry et al. (2019) document, IFM is an exercise in intercultural governance that requires carefully designed structures. Practitioners and academics emphasized that getting the governance right is dynamic and complex, but crucial for advancing IFM.

Concept #3: revitalizing Indigenous knowledge: Indigenous knowledge and practices are experiential and place-based

Indigenous peoples continue their fire management practices in many places. Sometimes there is an integration of Indigenous and Western knowledge systems in IFM programs. Where knowledge and practice have been discontinued, efforts to promote the revival of these practices are an urgent priority, and involve sharing knowledge across generations. Community goals for an IFM program in British Columbia, Canada included strengthening cultural connection and well-being, restoring the health of the land, and respecting traditional laws.

Fire knowledge is epistemologically rooted in a connection to place (Robinson et al. 2016). Fire management can be a spiritual ceremony, grounded in local stories or legends. IFM emphasizes a reciprocal relationship with the land, and a strong nexus between individual and landscape health. Under Indigenous epistemologies, the land often guides human action, while Western approaches are typically guided by politics, science, and economic incentives.

Drawing from Indigenous knowledge, IFM has been shown to be an effective place-based strategy to address wildfire and produce additional cobenefits and has important self-governance implications (Diver 2016). It is of critical importance that where Indigenous and Western knowledge are brought together, it is in ways that ensure respect, equity, agency, and decision-making power for all parties involved.

Concept #4: cobenefits: IFM generates multiple ecological, social, economic, and cultural benefits

In addition to reduced fire risk, literature demonstrates diverse market (income) and nonmarket (ecological, social, and cultural) outcomes from IFM, often termed cobenefits. IFM has generated significant greenhouse gas reductions by mitigating late dry season fires. The West Arnhem Land Fire Abatement (WALFA) project reduced carbon emissions from wildfire by 37.7% over seven years, relative to a preproject 10-year emissions baseline. The WALFA project also generates livelihood outcomes through carbon credit sales (Whitehead et al. 2008). Tribal nations in the United States are interested in generating livelihood opportunities from IFM (Rasmussen et al. 2007) and tribes in Northern California are implementing IFM programs (Diver 2016). IFM can also deliver "conservation-based development" (Rasmussen et al. 2007).

Individual participation of Indigenous peoples in environmental management (including fire management) has produced positive health outcomes, derived from enhanced physical activity, reduced psychological distress, and increased connection to the land (Burgess et al. 2005). Campbell et al. (2011:83) concluded these programs deliver "significant and substantial savings in primary health care expenditure for the management of chronic disease." IFM programs also deliver individual emotional and spiritual health benefits.

IFM programs enhance intergenerational knowledge and culture transfer, which is coupled with healthy and intact ecosystems (Fernández-Llamazares and Cabeza 2018). In Central Brazil, ceremonial practices and mentorship are combined to pass on fire knowledge (Welch and Coimbra 2019) and in British Columbia, educational programs and mentorship support IFM practices (Lewis et al. 2018).

"Pyrodiversity" describes the interactions between anthropogenic fire regimes, biodiversity, and ecosystem effects. In many savanna ecosystems, mosaic landscapes created through prescribed burning improve biodiversity. Landscape mosaics with many small fires result in more long-unburnt patches, which are important habitat for bird, reptile, and mammal species. However, Parr and Andersen (2006) argued the theory that "pyrodiversity begets biodiversity" has not received sufficient critical analysis, and that fire management should focus more on the specific levels of burning required for biodiversity goals, as well as operational guidelines for attaining them.

Concept #5: desired state of ecosystem: IFM can achieve a resilient ecosystem

Determining what is a healthy ecosystem is socially constructed, and rooted in worldviews, values, and experiences. Fire suppression in fire-prone regions has created a perception of safety, but has not achieved a desired healthy ecosystem state, and has paradoxically increased fire risk. Understanding cultural fire regimes can help restore fire-prone ecosystems and their services (Steen-Adams et al. 2019). The desired state of an ecosystem is evolving and subject to shifting baselines, where people's perception of a healthy ecosystem changes from one generation to the next, or even within one's own lifetime, as their outlook changes (Papworth et al. 2009). This presents a risk as landscape knowledge is passed down to younger generations through a lens where negative ecological changes may go unnoticed and unaddressed. IFM has the potential to address this lapse in ecological memory by encouraging intergenerational knowledge transfer on what is a healthy or resilient ecosystem.

The desire to achieve an ecosystem that is "natural" has epistemological flaws. The idea of "pristine nature" is a colonial idea that disregards Indigenous land management in markedly transforming the landscape. The concept of nature, land use, and Indigenous land justice are bound up in the politics of territorialization, which shapes norms and rules for legitimating land access and use. Further, climate change shapes what people see as the baseline of ecosystems, and hinders attempts to restore these to any baseline (Harris et al. 2006). Increasingly, there is a focus on integrating Indigenous knowledge and science with mainstream science concepts of systems thinking, resilience, and adaptive capacity in fire science, ecology, climate science, and many other relevant disciplines.

The need for adaptive strategies in fire management is critical. IFM must be specific to the ecological characteristics of the landscapes to which it is applied, and this varies across biomes, regions, and ecosystems. Contemporary IFM must also consider critical infrastructure, land tenure, public risk perception, and air quality, which are crucial when managing populated landscapes. There are multitudes of competing objectives, goals, and narratives that influence IFM and its objectives.

Conceptual model

The conceptual model identifies (A) multiple fire ontologies, with different ways of knowing and thinking about fire. At (1), the model shows these ontological differences create an epistemological divide bridged by collaborative governance, which influences, and is influenced by, the institutional context (B). At (2), the application of IFM brings this into a broader fire management regime (C). At (3), ontological differences affect the cobenefits people value and manage for (D), with the tendency in Western perspectives to favor market benefits. Ontology also influences our ideas of how a desired ecosystem should appear and function (E). At (4), these ideas will change as our baseline of what a healthy landscape looks like evolves. At (5), integrating Indigenous and Western knowledge systems through collaborative processes can transform fire governance, the methods used, and benefits produced, at (6). At (7), the cobenefits managed for will influence the desired state of an ecosystem, for instance, managing for timber or carbon will have different implications for the ecosystem than managing for biodiversity or cultural values. Balancing these different goals and values requires participatory and adaptive forms of management, and Indigenous knowledge and practices can support this. At (8), the link between desired ecosystem state and Indigenous knowledge points both ways, as Indigenous knowledge is often experiential and involves a continuous interplay between applying strategies and reading the landscape.

DISCUSSION AND CONCLUSIONS

IFM is a practice and tool that can bridge distinct epistemologies of fire (Mason et al. 2012, Lake and Christianson 2019). However, IFM is distinct and exists on its own terms, outside of Western fire management frameworks. The conceptual model presents five key IFM concepts from scholarly literature: fire ontologies, governance, Indigenous knowledge, cobenefits, and desired ecosystem state. The relationships between these are mapped out, illustrating the interdependencies between IFM concepts, as well as their accompanying ontological, epistemological, and methodological roles. These concepts play out to influence if and how IFM is used as a fire management strategy. This model aims to deepen understanding of the key elements of IFM represented in the literature, noting that most of this literature was developed by non-Indigenous academics. More effort to include and amplify Indigenous perspectives and worldviews on IFM is critical.

There are important temporal and spatial dimensions to IFM. The intervention depends on ecosystem type and season, and this creates challenges for knowledge sharing. Indigenous knowledge and practices are typically place-based and relevant to a specific fire regime and ecological context. For example, insights from north Australia's savannas are not readily transferable to Canada's boreal forest. That being said, the main concepts illustrated in the conceptual model were found to be representative of global IFM literature. The conceptual model captures important commonalities to implementing IFM.

As societies and communities learn to coexist with fire and expand thinking beyond Western ontologies and perspectives, there are important issues to consider. If IFM is pursued through market-based mechanisms (like carbon credits), will this institutionalize and constrain the dynamism and experiential nature of Indigenous fire knowledge? Reciprocal relationships, and the connection between physical and spiritual worlds, are not easily integrated into Western scientific approaches (if integration is a management goal). New forms of governance will be important, and while adaptive comanagement is one way for bridging epistemologies and building trust, there remains questions about the ownership of Indigenous knowledge. One fire academic observed, "academics come and take the stories [of cultural fire] and it gets put out into the academic world, and that knowledge is taken away from the community." Putting safeguards in place to preserve Indigenous ownership of knowledge is crucial.

Institutions and power dynamics will shape the evolution of IFM. An Indigenous fire practitioner stated, "for Indigenous communities there is both a cultural and an ecological risk to *not* burning" [emphasis added]. They elaborated that "[t]his results in risks for Indigenous communities performing their culture. Should they risk arrest to perform cultural fire ceremonies? Or should they risk a cultural loss by not performing them?" These laws, norms, and power dynamics may shift over time as IFM is implemented and legitimated by the dominant regime. There remain important questions as to whether IFM will expand beyond Indigenous land tenures, into contested areas as crises accelerate the acceptance of other fire solutions. As a fire academic reflected, "IFM is being considered as a solution because megafires are occurring, and this leads to a paradigm shift in fire management." It could be that as the benefits of IFM are demonstrated, this could facilitate a paradigm shift away from fire suppression, to more proactive and community-driven approaches. Further research is needed to better understand whether there exists a transformation toward IFM because fire suppression is demonstrably ineffective, or whether there is a broader shift in ontologies and epistemologies recognizing the importance of Indigenous knowledge for sustainable development in a changing climate.

Social forestry in India

Figure: Social forestry near Mothugudem of Khammam district in Andhra Pradesh, India

Social forestry is the management and protection of forests and afforestation of barren and deforested lands with the purpose of helping environmental, social and rural development. The term social forestry was first used in 1976 by The National Commission on Agriculture, when the government of India aimed to reduce pressure on forests by planting trees on all unused and fallow lands. It was intended as a democratic approach to forest conservation and usage, maximizing land utilization for multiple purposes.

The Indian government attempted to expand forest areas which were close to human settlements and which had degraded due to human activities. Trees were planted along railway lines, roadsides, rivers and canal banks, in village common land, government wasteland, and *panchayat* land, and were to be planted in and around agricultural fields. Among the goals were to increase fuel availability in rural areas and to prevent soil erosion. This program was a failure due to the lack of governance, and management was delegated to the village *panchayats* (village councils).

Demands and development

People felt the need for a social forestry scheme because India has a dominant rural population that depends largely on fuelwood and other biomass for their cooking and heating. These demands will increase with population, threatening to reduce forested areas.

Participation

The social forestry scheme sought to enable the common people to raise plantations which would meet the growing demand for timber, fuelwood, fodder, etc., and thereby reduce pressure on traditional forest areas. This concept of village forests to meet the needs of rural people had existed for centuries across India. With the introduction of this scheme, the government formally recognized the rights of local communities to forest resources and encouraged rural participation in the management of natural resources. Through the social forestry scheme, the government has involved community participation, as part of a drive towards afforestation, and rehabilitating the degraded forest and common lands.

Bihar, one of the poorest states of India, lacked National Rural Employment Guarantee Act (NREGA) employment during the flood season, and this work was only suitable for able-bodied people. Linking the social forestry scheme to NREGA was hoped to reduce poverty and climate change. To ensure adequate care for the plants, their ownership was given to those who had cared for them for five years. Within 3 years, forest coverage in the region grew from 7% to 12.86%, and provided employment to thousands of women, handicapped and elderly people.

After managing the projects for five years, the government delegated authority to the village *panchayats* (village councils) to manage for themselves and generate products or revenue as they saw fit.

Objectives

Social forestry schemes have the main objectives to:

- Improve the environment for protecting agriculture from adverse climatic factors,
- Increase the supply of fuelwood for domestic use, small timber for rural housing, fodder for livestock, and minor forest produce for local industries,
- Increase the natural beauty of the landscape; create recreational forests for the benefit of rural and urban populations,
- Provide jobs for unskilled workers,
- Effect land rehabilitation, and
- Raise the standard of living and quality of life of rural and urban people.

The mission of the scheme is:

- To carry out a need-based and time-bound program of afforestation with special emphasis on fuelwood and fodder development on all degraded and denuded lands/forests.
- Effect afforestation of abandoned jhum lands and mined areas.
- Establish linear strip plantation of fast-growing species on sides of public roads, rivers, streams and irrigation canals.
- Effect afforestation on under-utilized lands under state, institutional or private ownership.
- Create green belts in urban/industrial areas.
- Create shelterbelts (generally more extensive than the windbreaks) for the purpose of shelter from wind and sun covering areas larger than a single farm on a planned pattern.
- Establish farm forestry in the form of raising rows of trees on boundaries of fields and individual trees in private agricultural land as well as the creation of windbreaks around a farm or orchard by raising one or two lines of trees.
- Raise flowering trees and shrubs to serve as recreation forests for the urban and rural population.
- Elicit people's participation, involving women and young people in the conservation of forests, wildlife, and the environment.

- Generate environmental awareness and promote environmental events

Types

Social forestry schemes can be categorized into groups: farm forestry, community forestry, extension forestry and agroforestry.

Farm forestry

In Farm forestry or agroforestry, trees are grown on farmland for commercial and non-commercial purposes. Farmers are encouraged to plant trees on their own farmland to meet their domestic needs. A tradition of growing trees on farmland already existed in many areas, and was the main thrust of most of India's social forestry projects. In addition to providing fuelwood, farmers often grow trees to provide shade for agricultural crops, as wind shelters, for soil conservation or to recover wasteland.

The pulp and paper industry is a major demand driver for certain species of tree such as *Eucalyptus*, *Babul Acacia catechu*, *Subabul* (*Leucaena leucocephala*) and *Casuarina equisetifolia*. As a rough estimate, the total demand for pulpwood is approximately 10 million ADMT (wood having 10% moisture). The Indian Paper Manufacturer's Association is an umbrella organization that coordinates and drives plantation efforts by member organizations in India to supply its industrial requirements.

Community forestry

The government provides seedlings and fertilizer to the community, which is then responsible for nurturing and protecting the plants on community land. This provides for the community rather than individuals. Fast-growing *Eucalyptus* has been planted on a large scale. Some communities manage these plantations sensibly and in a sustainable manner for continual benefit, while others sell the mature timber for a one-time capital divestment.

Extension forestry

Planting of trees on the sides of roads, canals and railways, along with planting on wastelands is known as extension forestry, increasing the boundaries of forests. This has created wood lots in the village common lands, government wastelands, and *panchayat* lands.

Agroforestry

In agroforestry, silvicultural practices are combined with agricultural crops like legumes, along with orchard farming and livestock ranching on the same piece of land. It is defined as a sustainable land-use system that maintains or increases the total yield by combining food crop together with forest tree and livestock ranching on the same unit of land, using management practices that consider the social and cultural characteristics of the local people and the economic and ecological condition of the area.

Partners

Major non-governmental organizations who partner in the implementation of the social forestry scheme are:

- Sri Soneswar Nath Mahadev Trust
- Try
- National Green Highway Mission (NGHM)
- Bihar forestry Development Corporation Limited.